CHOSEN PEOPLE FROM THE CAUCASUS

CHOSEN PEOPLE FROM THE CAUCASUS

Jewish Origins, Delusions, Deceptions and Historical Role in the Slave Trade, Genocide and Cultural Colonization

Michael Bradley

THIRD WORLD PRESS
Chicago

CHOSEN PEOPLE FROM THE CAUCASUS
Jewish Origins, Delusions, Deceptions and Historical Role in the Slave Trade, Genocide and Cultural Colonization

by Michael Bradley

THIRD WORLD PRESS
CHICAGO, ILLINOIS

First Edition 1992
First Printing 1992
ISBN: 0-88378-147-6

Cover design by Craig A. Taylor

Manufactured in the United States of America
Third World Press
7822 So. Dobson Ave.
Chicago, Il 60619

Dedication

For Winnie and Jason,
George and Ellen.

CONTENTS

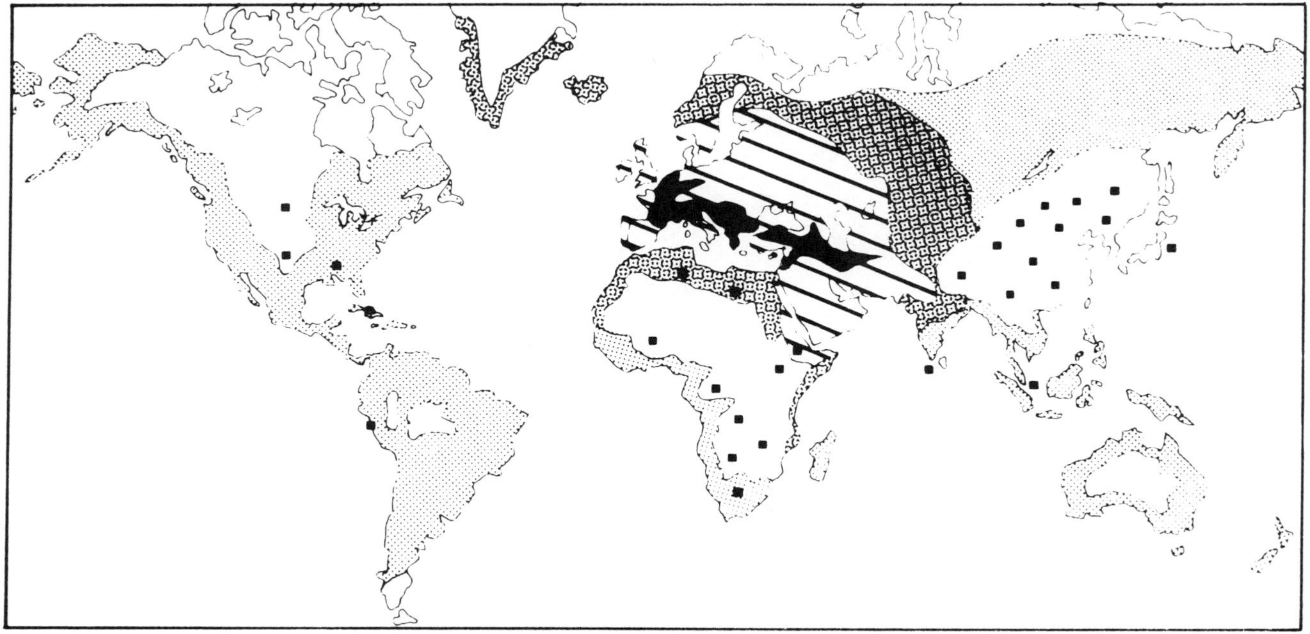

The Iceman Inheritance:

The Expansion of Neanderthal-Caucasoid 'Western' Influence

- Neanderthal homeland 100,000 - 30,000 B.C.
- Neanderthal-Caucasoid expansion 30,000 - 1,500 B.C.
- Euro-Arabic 'Western' expansion 1,500 B.C. - A.D. 1,400
- European colonial expansion A.D. 1,400 - 1850
- 'Western' colonial, technological cultural expansion A.D. 1850 - present
- Modern resistence to Western expansion and colonialism based upon borrowed Western cultural, political, technological developments

FOREWORD

In his previous books, *The Iceman Inheritance* (1978), *Dawn Voyage: The Black African Discovery of America* (1980) and *The Columbus Conspiracy* (1991), Michael Bradley has opened some long-closed historical doors. In doing so, he has disturbed some well-preserved skeletons and frightened a lot of living people. He has dared to explore some of the side streets of history and unearthed long-buried truths from their academic resting place. Whereas for the last 500 years, most historical records have been distorted to make people of European descent feel good at the expense of the majority of mankind, Michael Bradley's ideas have made people of European descent feel restless and somewhat betrayed because he is also "one of them."

He grew up in the American South, an atmosphere that normally would have prepared him to defend and protect the myths of history that he has exposed. However, he has emerged as a "born again seeker of truth" about the human condition.

In the Introduction to *Chosen People From The Caucasus*, Michael Bradley explains his reasons for writing his previous book, *The Iceman Inheritance.* He also explains how and why the published version of *The Iceman Inheritance* fell short of his original intent. The book was a broad approach to the subject of racism and how it affects the lives of most European people. A number of people who had previously avoided the subject of racism were forced to face it frontally for the first time in Bradley's work, where the author boldly stated that the book was racist. If the book is properly understood, it can serve as a significant force in dismantling the concept of racism.

In *Chosen People From The Caucasus*, Michael Bradley again opens forbidden doors and announces his findings. He has already proven that he is an astute and restless research scholar who goes beyond the norm in search of the hidden truths behind subjects that are generally spoken about in whispers, if they are spoken about at all.

During the period of slavery, Africans in America estranged from their culture and their religion wanted to associate with a people who had escaped from bondage. Those who could read the Bible accepted the story of the Exodus literally. This story of Jews escaping from bondage in Egypt became more real to them than to others because they were slaves holding onto the thin hope of escape. Jewish folklore became real to them. This is a part of Black America's sentimental attachment to the Jewish people, an attachment that is sometimes called friendship and sometimes called an alliance. It is in actuality an indulgence in fantasy and in biblical folklore.

Today, a large number of Black Americans think of the Jews as people of The Book, and they look more favorably on them than on other Caucasians. These poorly informed Blacks do not know that the Jews of Europe played the same role in their enslavement, oppression and colonization as other Europeans. This is why *Chosen People From The Caucasus* will cause heads to shake, change many minds and challenge people to discard their long-held fantasy about the Jewish people of Europe and the assumption that they were friendly toward African people over and above other Europeans.

In Chapter One, Michael Bradley points to the fact that:

> As financiers and insurers of European maritime commerce, a commerce largely focused on the slave trade for three centuries, Jewish mercantile houses amassed vast fortunes in both Europe and North America, fortunes that were often inter-related by family and marital ties. In Europe, "conversos" operating out of Madrid and Seville handled the business of Spanish slaving; Jews openly operating in Amsterdam served the financial needs of Protestant slavers. In America, a flourishing Jewish financial community centered mainly in Newport, Rhode Island, financed and insured slaving voyages by Yankee captains.

He also highlights the disproportionate role that Jews have played in relation to the spread of what is called Western civilization. After having been pushed into trades and professions other people did not want or considered to be beneath their social status, many Jews became experts in these professions that eventually became indispensable to the State. The paradox here is that many people who had originally forced them into these professions, later condemned them for using these professions.

For example, their skill in banking and other forms of high finance became a trademark that made some people both respect and fear them. In examining the influence of Jews in medicine, psychiatry and the mass media, we learn that Jews have "built an empire of their own" in more ways than one. Because they are influential in these essential areas of communication, healing and the mind sciences, their influence extends far beyond their numerical presence in the population. When one calls public attention to this fact, he/she is generally labeled as being anti-Semitic. This charge, too often repeated, tends to infer that Jews are not to be judged by the same rules as are the rest of humanity. At this point we should discuss the chosen peoples' concept of self and ask, "Who did the choosing, and what were they chosen to do?"

It is not easy to understand the role of the people referred to as Jews in human history without a serious attempt to locate them on the map of human geography. What, exactly, is a Jew? How do the people referred to as Jews relate to the biblical people of the Hebrew faith, if at all?

The people we now confront and recognize as Jews are of European origin. At once this tells us something about the will of Europeans to dominate and interpret the cultures and concepts that govern the world. When we speak of those Europeans who refer to themselves as Jews, we are dealing with a supersensitive area where most people will hear what you did not say while stubbornly refusing to hear what you actually said.

The word "Jew" is a European creation. Yet, there are people of the Hebrew faith all over the world, of many different colors and many different cultures. Dr. Y.A.A. ben-Jochannan's *Welcome Back Home,* as well as his *We the Black Jews: Witness to the "White Jewish Race" Myth* bring to light the universality of people adhering to the Hebrew faith, long before the belated conversion of those Europeans who now dominate the faith and assume that they must give it its world definition and direction.

Chapter Five deals with the Khazar conversion of the European Jews, a neglected chapter in their history. Here, Bradley reveals that the formation of the people referred to as Jews in Europe was not the same as the origin and development of a people in western Asia who bore the same name. This is a clear indication that a variety of people of different cultures, origins and colors became followers of the Hebrew faith and are historically known as Jews, but do not share the same history as the Jews depicted in the *Old Testament.*

The history of the Jewish people is shrouded in myth and mystery. They have played a role in the rise and fall of many nations and have had an influence on nations and cultures far beyond their numbers. Both literally and figuratively, the Jews are a people under siege and on the offensive. In many cases, their survival has depended on their knowledge of the prevailing power apparatus of the countries where they were either permanent or temporary residents. Their role in the Atlantic slave trade is unique because at the time they were playing this role, they were also fighting against the possibility of being enslaved themselves. Likewise, they have learned to profit from both war and peace depending on the circumstances.

Chapter Six, "Delusions, Deceptions and Distortions," is the most valuable for the modern reader because it is a detailed account of the participation of the Jews in the Mediterranean world and their activities that made the European renaissance possible. This renaissance introduced Europe to the maritime skill and the revolution in navigation and boat-building that eventually led to the slave trade that was a three-continent industry. This period subsequently became a pivotal turning point in history. During the 15th and 16th centuries, Europeans would not only colonize most of the world, they would colonize most of the information about the world; the African Holocaust, called slavery, was set in motion; the labor and resources of Africa and parts of Asia would lay the basis for the modern scientific and technical world. Jews were major players as well as victims in this essential turning point in world history.

It is clear after reading this book that Michael Bradley will not be a candidate for a teaching position in any American or Canadian university. His is a courageous voice in a wilderness of misconceptions. The serious listeners to his message will grow very slowly, but they *will* grow and reward him for his efforts to open long-closed doors of information.

John Henrik Clarke
Professor Emeritus
African World History
Hunter College
New York
April 1991

INTRODUCTION

In *The Iceman Inheritance*, I offered evidence from recent geological research, physical anthropology and history that Caucasians are the most aggressive major genetic group of humanity. In 1978, such a notion seemed an affront to both common sense and to the sensibilities of North America's Caucasian majority, therefore I felt forced to open the book dramatically:

> This book is racist.
>
> For, among other things, I will attempt to show that racism itself is a predisposition of but one race of Mankind—the white race.
>
> I believe that I can show that our converging

> contemporary crises, like racism itself, have their origins in the prehistory of the white race alone. We attribute various threats to our survival to Man's folly, but this is a conscious and self-protecting euphemism.
>
> Nuclear war, environmental pollution, resource rape... all are primary threats to our survival and all are the result of peculiarly Caucasoid behavior, Caucasoid values and Caucasoid psychology.
>
> There is no way to avoid the truth. The problem with the world is white men.
>
> If this insight has generally escaped us, it is only because we have dictated much history, written most of it, and have judged it in terms of our own self-image. But we have never looked at ourselves in an objective biological mirror.
>
> And that is what we will now proceed to do.[1]

The next 226 pages of *The Iceman Inheritance* argued that Caucasoid ancestors and their modern Caucasian descendants are significantly more aggressive than other major genetic groups because of their uniquely glacial evolution. Citing evidence from physical anthropology primarily, I suggested that purely physical adaptations demanded by survival under arctic conditions during the last (Wurm)[2] Ice Age in Europe conflicted with, and partly erased, previous adaptations designed to absorb and displace aggression. The result of this progressive glacial adaptation was that Neanderthal-Caucasoids became increasingly maladapted psychosexually, suffered extremes of psychosexual ambivalence and frustration and expressed the resultant aggression to a greater extent than peoples of other major genetic groups. Apart from sheer violence, I tried to show that inordinate Caucasoid psychosexual frustration and aggression were revealed in almost every aspect of Western culture: geographical expansion at the expense of other peoples and subjugation of the conquered; chauvinistic, intolerant and restrictive religious conceptions and practices that were established as Western cultural norms and then imposed, by force, upon non-Western subjects; extraordinary sexual differences in fashion; emphasis upon "love" in art and literature; the dichotomy between carnal and spiritual love; restrictive and anti-feminist aspects of "love" itself; unique development of technology to an environmentally unsupportable level; the unique penchant for applying technology immediately to warfare; and much more.

The Iceman Inheritance was intended to be a very serious book, the result of fourteen years of research—but it was compromised by many flaws. First and foremost among them were my own grave limitations of scholarship; I tried to cover too much of a subject that was, in any case, much too vast to be adequately encompassed by one mind—a mind that had been self-educated through force of circumstance. Additionally, because of the book's controversial nature and the author's lack of formal academic credentials, *The Iceman Inheritance* was shunned by established and experienced publishers who might have offered valuable assistance in the proper presentation of provocative and potentially valuable ideas. Instead, after many rejections, the manuscript was finally accepted by a fledgling, would-be Canadian publisher who wanted *The Iceman Inheritance* for his very first book.

Purely financial considerations dictated a significant

reduction in length and much supporting research was edited out, in haste, just before the book went to press. In addition, Dorset Publishing Inc. insisted on a "popular" book without academic footnotes, so the lengthy references were also cut. And, most unfortunately, two subsequent editions of *The Iceman Inheritance* were merely reprints of the original Canadian typesetting. Thus, all the flaws have been perpetuated for an ever-increasing readership.[3]

Nonetheless, the basic argument of *The Iceman Inheritance* has been supported, in the fourteen years since its original release (November 1978), by scientists and researchers with the credentials that I lacked. The first "big gun" support came just one month after Canadian publication. Dr. Eric Trinkhaus (then of Harvard) and Dr. T. Dale Stewart (Smithsonian) jointly published "Hard Times Among The Neanderthals" in the December 1978 issue of the American Museum of Natural History's prestigious *Natural History* magazine. Trinkhaus and Stewart subjected all known Neanderthal skeletal material to a coroner's style forensic examination. Their conclusion was that Neanderthal bones exhibited a much greater number of violence-inflicted injuries than any comparable number of bones from other groups of human ancestors. Since it had long been known, and accepted, that the Neanderthals were one ancestral group that contributed to the modern European gene pool, it seemed reasonable to conclude that this proven Neanderthal aggression had been passed on to their modern European descendants. This is what I argued in *The Iceman Inheritance*, but had argued from what might be called a "philosophical biology" perspective; Trinkhaus and Stewart provided the hard data supporting an important part of my theory.

In the early 1980's, a West German researcher named Ilse Grubrich-Simitis was working on a book about the career of Sandor Ferenczi[4] when, in a trunk full of Ferenczi's paper and correspondence, she came across a 12-page tract in Sigmund Freud's handwriting. Freud had written this essay in 1919, but its existence had been unsuspected for over sixty years. In this newly-discovered essay, Sigmund Freud came perilously close to overturning his own theories of psychosexuality and psychotherapy because he concluded that many, or even most, acute anxiety neuroses stemmed not from childhood trauma, but from racially remembered centuries-long trauma of trying to survive during the last Ice Age. According to this short, 12-page essay, survival outside the stone age caves was so brutal that the only respite from it were "perverse" sexual pleasures within the caves where Ice Age men took their frustration, aggression and fears out on their women.

Freud's brief essay was published, with copious notes and comments, by Harvard University Press in April 1987 (edited by Ilse Grubrich-Simitis, translation and foreword by Axel and Peter Hoffer) under the title *A Phylogenetic Fantasy*. Discounting the usual Freudian lurid imagination and exaggeration, *A Phylogenetic Fantasy* argued precisely what *The Iceman Inheritance* had argued nine years earlier, but on considerably more scientific grounds than Freud: that much of "our" behavior was conditioned by an Ice Age evolutionary experience, behavior still genetically transmitted and/or selected by cultural values of Ice Age profile, and that the essence of this aggressive behavior was psychosexual maladaptation. Thus, *The Iceman Inheritance* came to be supported by a lost work of none other than Sigmund Freud.

This 12-page paper is most interesting in suggesting, or *asserting*, that an individual's behavior in the 20th century can be explained by cultural perpetuation of Ice Age behavior-patterns supported to an unknown extent by actual biologically transmitted genetic traits. This is the position defended in *The Iceman Inheritance*. For those unwilling to grapple with Freud's own purple prose, much less the turgid commentary of Grubrich-Simitis, David Cohen's "Hangups Due to Ice and Id of Bygone Era" in the August 6, 1987 issue of the *London Observer* gives a good idea of Freud's lost theory. Another popular presentation, by Dr. George Steiner of the University of Geneva, can be found in the *Sunday Times* (*London*) issue of September 6, 1987.

Personally, I am not at all sure that Freud's 12-page essay was legitimately "lost"; I am inclined to the suspicion that it may have been *hidden*. Freud believed that Ice Age experiences had a profound influence upon present-day "hangups." However, by about 1919, it was known that the so-called Ice Age had affected only Europe and part of western Asia so far as human evolution was concerned.[5] Ancestral human populations in Africa, the Far East and in Indonesia-Australia were unaffected by the Eurasian cold. In these tropics, the only effect was a slightly wetter climate, the "pluvials." If Freud was forced to conclude that Ice Age cold traumatized people, then he could not escape the truth that only *one kind of humanity* evolved in a glacial environment. That expression of humanity were the Neanderthal-Caucasoids of Europe and western Asia, and them alone. The ancestors of modern Western man. Was Freud prepared to reveal that only Western mankind remained psychosexually traumatized? Was Freud prepared to admit that his grandiose construct of psychotherapy simply did not relate to non-Western humanity? Evidently not.

Aside from the forensic work of Trinkhaus and Stewart, and in addition to Freud's lost manuscript, other researchers working along independent lines were coming up with theories that paralleled *The Iceman Inheritance* without contradicting it. As early as 1970, Dr. Frances Cress Welsing had proposed that the presence or absence of melanin might affect more than skin color; it might affect perception and behavior. She has expanded on this idea in *The Isis Papers: The Keys To The Colors.*[6]

If Caucasoid humanity is, indeed, somewhat more aggressive than other major genetic groups, then it stands to reason that truly civilizing influences must have come to Western man from other peoples. Cheikh Anta Diop argued just this in his *Civilization or Barbarism: An Authentic Anthropology,*[7] while more recently Martin Bernal, a white Jewish scholar at Cornell, has shocked the academic world with his *Black Athena: The Afroasiatic Roots of Classical Civilization.*[8]

As of this date, the amount of converging and accumulating data has, by any reasonable criterion, lifted the arguments of *The Iceman Inheritance* well beyond the status of theory into the realm of being a working hypothesis—a conjecture that accounts for a set of facts and which can be used as a basis for further investigation. *Chosen People From The Caucasus* continues my investigation of Neanderthal-Caucasoids to discuss the origin and historical role of the "Hebrews" and their participation in the slave trade, genocide and cultural colonization.

I
The Chosen People

Before beginning this investigation, some clarifications and definitions are in order. Hopefully, the following points will serve to prevent sincere and not-so-sincere misunderstandings, misrepresentations, misapprehensions, misconceptions and misquotations of this book.

1. *The Iceman Inheritance* does not suggest that Caucasians are the only aggressive major group of humanity. All of humanity is highly aggressive, and all of humanity exhibits psychosexual confusions, frustrations and violence to a degree not even dimly reflected among most other animal species (except among the species closest to man, the ground-living drills and baboons, and the four known kinds of anthropoid apes). All human beings suffer

from psychosexual maladaptation, and all human beings have proved to be highly aggressive in consequence. However, in *The Cronos Complex*[1] and *The Iceman Inheritance,* I argued that Caucasians seem to be significantly *more aggressive* than other major groups of humanity and this fact has had an identifiable impact upon human history as well as molded the norms and values of our present Caucasoid-imposed world culture. Western man, however, has no monopoly on aggression, no monopoly on psychosexual disturbance.

2. I was justly criticized (by Dr. Judith Posner, York University, Toronto)[2] for using the term *race* in *The Iceman Inheritance* without defining it, and without referring to definitions of others. Because of the emotionalism attached to the word "race,"[3] Ashley Montagu has suggested the alternative term, "major groups"—but again, without defining them. Already, in these few pages, the term *major genetic group* has been used as a kind of synonym for *race*. The author presents the following definitions for readers of *The Iceman Inheritance* and this book:

> Race/Major Genetic Group/Major Group—a human population, usually before the 16th century characteristic of a geographic region, which, by genetic amalgamation or genetic isolation developed visually obvious physical characteristics (and, perhaps, not visually obvious physio-psychological characteristics) distinguishing it from other major genetic groups.

In accord with this definition, Montagu would accept the existence of four *races* of mankind (which he calls "major groups"): the "Negro" or "Black," originally domiciled in Africa; the "Caucasian" or "White," originally domiciled in Europe but represented as far as western Asia; the "Mongoloid" or "Yellow-Brown," originally domiciled in eastern Asia and separated from the Caucasians by the Pamir-Tien Shan-Altai mountain barrier which trends from the southwest-to-northeast through Central Asia; and the "Australoid" or "Archaic White" which now inhabits only the outback of Australia, but which once domiciled itself in Southeast Asia-Indonesia-Australia.[4]

To this list of four "major groups" of humanity, Carleton Coon would add one more: the "Capoids" or Bushmen, a race also originally domiciled in Africa, but earlier and distinct from the later Negroids or Blacks, having very marked physical-sexual traits not possessed by other major groups and some other characteristics similar to Mongoloids.[5]

Other physical anthropologists have proposed other races as well, most estimates ranging from "six to nine,"[6] but, when examined, these additional racial candidates reflect uncertainty about the status of pygmy peoples that otherwise resemble full-size neighboring populations living in "non-marsh" nearby habitats. Veddahs are pygmy Caucasians living in marshy places in Sri Lanka; Congo pygmies are small people living in the Congo river basin but otherwise similar to neighboring Africans; the *orang pendek* of Java and Sumatra appear to be small-size versions of the nearby Mongoloids. But the "racial" and evolutionary status of all these pygmies is in doubt. Experts do not agree. The author's definition of Cultural/Ethnic group is:

> a human population, belonging to *one* of the aforementioned racial/major groups, which, through a certain amount of geographic isolation, has developed

> certain social values and behaviors which make it distinctive.

For example, Sioux Indians in North America, Japanese and Chinese are all of one *racial* or *major group* of humanity, but have very different cultural or ethnic expressions due to migration into different geographic areas and subsequent isolation. The author's definition for Linguistic Group is:

> a human population which speaks a certain language due to recent or ancient adoption of that language. This is *not* any indication of *ethnic*, much less *racial*, affiliation.

All of this can be so easily muddled, and muddied, as we shall see when we come to our discussion of the Hebrew "race," "ethnic group" and Semitic *language*. In Canada, there is a contemporary issue of "racial conflict" between French-speaking Canadians and English-speaking Canadians. There are charges and countercharges about "racial discrimination" between the two groups. Even politicians and educators (who, presumably, should know better) have referred to "racial" and "racist" problems between Canadian Anglophones and Canadian Francophones—and yet it is all absurd. The French-speaking Canadian settlers and the English-speaking ones, not only come from the same *race* or "major genetic group," as defined above, but share a Western European cultural and ethnic tradition; even their respective languages are so closely related that they are classified within the same branch of the Indo-European linguistic family: on a practical level, this means that very basic French, if spoken slowly, can be understood, more or less, by an English-speaker, and vice versa. The Canadian conflict is over language and nothing else, not culture, and certainly not race.[7] This example may be helpful when we come to the slightly more muddled (and purposefully muddled) matter of Jewish racial, cultural and linguistic affiliations.

It is not very difficult to learn another language well enough to get along in any ordinary conversation, although Americans seem most unwilling to do this. In Europe, it is not unusual for an individual to speak, read and write three or four languages with proficiency. It is much more difficult to acquire another *culture*. One must almost be born into a culture, and grow up with it, in order to absorb all the numberless and subtle values, expectations and behaviors comprising it—and these go far beyond mere language. Some Europeans, Africans and Asians are truly "multi-cultural" so long as they stick to cultures developed within their own *race*. An adept Japanese can truly absorb, say, Korean culture so that as long as he speaks faultless Korean, no one would be able to guess that he was originally from Japan. The same holds true for a Czech, say, living in Poland, a Frenchman living in Scotland, an Ibo living among the Anyang.

But it may be doubted if anyone can become truly "multi-racial" in the same sense, even if superficial physical differences are somehow ignored in day-to-day interactions and even if the local language and culture are mastered adequately. There almost always *must* be some subtle differences in perception and conception of reality that no interloper can master completely. That is why, in any cosmopolitan city, people of various *racial* groups almost invariably gravitate to the same urban area, forming a "ghetto." People belonging to highly distinctive *ethnic/ cultural groups* tend to do the same thing. They form a distinctive enclave where they can "feel more at home," and understand their neighbors in a way that goes far beyond language or even culture. The phrase "highly distinctive culture" means, almost by definition, a group that has experienced significant historical isolation, imposed either by geography or by social pressures (or both), so that traditional costumes, hairstyles, linguistic dialects, etc., have been anachronistically retained. These *non-racial* physical distinctions not only set them apart from their racial fellows, but form an important buttress to group and individual identity—particularly for the first few generations after relocation into a more cosmopolitan society. Toronto, a city that has supported "multiculturalism," has even manufactured bilingual street signs for obviously ethnic areas: Greeks gravitate to the east-central part of the city, and the street signs are in Greek and English; Italians are concentrated in the west or central part of the city, "Little Italy," and the street signs are in Italian and English; Jews are settled in the north-central part of the city, but Yiddish is not used on street signs; while various Slavic groups have gravitated toward the west end. These distinct ethnic, cultural (and linguistic) groups all belong to the same *race*—the Caucasian.

Toronto, like other Western cities, does not know what to do about "race." Toronto has a "Chinatown" where the street signs are in Cantonese and English, but all the Orientals are lumped together without regard for the very distinctive Chinese *ethnic group*, the equally distinctive Vietnamese *ethnic group* and the extremely distinctive Japanese *ethnic group*—which are all members of the

Mongoloid *race*. These Far Eastern ethnic and cultural groups do not have street signs in their own languages (Europeans have been favored again!), and maybe someone should agitate for this fair play.

Sometimes, a transplanted ethnic and cultural group becomes so numerous and so concentrated that, within its ghetto, an individual can be self-sufficient without having to assimilate. In Toronto's "Little Greece" and "Chinatown," generations of people have lived without learning English and without assimilating into Canadian culture.

In North America we are witnessing a similar but slightly different phenomenon; a *racial* group, "African-Americans," has become so numerous that it is starting to influence the nominally Caucasian mainstream culture. Dr. Ray Winbush, Director of the Bishop Johnson Black Cultural Center at Vanderbilt University, commented: "America's so-called youth culture *is* Black Culture." Trends in fashion, gestures, slang and music come to mind and one must agree that this is true. Particularly consider the phenomenon of "rapping." It *might be* (notice the qualifying words) a true instance of *racial*, as opposed to ethnic and cultural, proclivity re-asserting itself in an unlikely situation, perhaps a symptom of African emergence from the Caucasian society into which they were so brutally forced.

Anyone with even a superficial knowledge of African languages knows that they are melodious and rhythmic. Africans always took a delight in verbal virtuosity. In remote parts of West Africa, even as late as the 1930's, white visitors witnessed African lawsuits argued in rapid and unbelievably clever extemporaneous verse.[8] African languages are made for "rapping." But, the vast majority of African-Americans have never spoken any language other than English, which is not suited for "rapping." Nonetheless, contemporary African Americans have managed to revive a traditional African and *racial* trait, which most of them could never have known about consciously, and have figured an unsuitable language into the bargain. Do African Americans have a greater gift for rapping than other people, and if so, is that evidence of a *racial* capability not shared to the same degree by representatives of other major genetic groups? Popularly, it is considered that although whites *can* learn to rap, they are not as good at it as Blacks, on average. "Oh, he's not bad . . . for a *white* rapper," is the usual recognition of a praiseworthy attempt!

These two long diversions clarify the notion that while Caucasians seem to be *more aggressive* than other major groups of humanity, all races of man have exhibited a high degree of psychosexual frustration and aggression. Caucasians have no monopoly on violence; and the definitions of *race, cultural/ethnic group* and *linguistic group* should prove helpful in our discussion.

If the evidence seems to compel the conclusion that the Caucasian *race* is somewhat more aggressive than other major groups of humanity, then we may legitimately inquire as to whether some *ethnic* groups within the Caucasian *race* might be characterized by an even higher level of psychosexual maladaptation and therefore have historically manifested a consistently higher level of aggression than most of the Caucasians. Just as Caucasians have come to dominate the world by expansion and conquest, by technology and cultural imposition, is there a Caucasian group that has come to dominate Caucasian civilization? Jews immediately spring to mind. Given their relatively small numbers, Jews have had an inordinately great influence on Western civilization. And, given the fact that Western civilization has come to dominate the world over the past five hundred years, Jews have accordingly played a disproportionately major role in the unfolding drama of all humanity.

Jewish historical and religious experience in the ancient Middle East, related in the Old Testament and supplementary rabbinical texts, actually laid the foundation of the West's three major religions: Judaism, Christianity and Islam. So, it can be said with no exaggeration at all that Jews significantly molded the spiritual orientation of the entire Western world.

Jews had immense influence in the expansion of Europe that began in 1492 and culminated in world domination. Four "conversos," Jews who had supposedly and voluntarily converted to Roman Catholicism, put up most of the money to finance the first voyage of Christopher Columbus.[9] In the early days of the conquest of the Americas, from 1494 to about 1550, Jews were a numerous minority, if not an actual majority, of the so-called "Spanish" colonists that settled the New World.[10] Many of the "conquistadors" were Jewish[11] and, as such, took part in the most brutal episode of genocide and extermination the world has ever witnessed: within a century of 1492, the Native American civilizations had been destroyed and the population decimated by enslavement, starvation and disease. When Native American labor grew scarce because of genocide, Africans were forced into the New World tropics to slave in the mines and on plantations. It has been estimated reliably that between ten and fifteen *million* Africans and American Indians perished in slavery between 1494 and 1865.[12] But some estimates range as high as 100 *million* victims.[13] As conquistadors, financiers and insurers of slaving voyages, and as substantial slaveowners in the West Indies and Caribbean, Jews played a significant role in this infamy—a role that comtemporary Jews are prone to deny, disguise and distort in favor of their own victimization in Nazi Germany 450 years later.

As financiers and insurers of European maritime commerce, a commerce largely focused on the slave trade for three centuries, Jewish mercantile houses amassed vast fortunes in both Europe and North America, fortunes that were often interrelated by family and marital ties. In Europe, "conversos" operating out of Madrid and Seville handled the business of Spanish slaving. Jews openly operating in Amsterdam served the financial needs of Protestant slavers. In America, a flourishing Jewish financial community centered mainly in Newport, Rhode Island, financed and insured slaving voyages by Yankee captains.[14]

It cannot be said that Jews were responsible for the transatlantic slave trade, nor were they the only people involved in and profiting from it, but Jews were undeniably major players. Most transatlantic commerce of the 15th-19th centuries, even when they were not actual voyages to capture and transport Africans as slaves, involved the products made possible by slave labor: cotton, sugar, molasses, tropical woods and rum (all produced on plantations using field slaves). Gold, silver, gems and other minerals were also produced by slaves worked to death in New World mines.

One way or another, the great fortunes of Renaissance Europe were built on this squalid foundation, and the Jewish fortunes were no exception. This vast wealth went to establish the famous Jewish banking houses of the 19th century: Rothschild, Kuhn, Loeb, Warburg and others. Interconnected by marriages and familial alliances, these Jewish financial interests, collectively, came to have immense influence on the fiscal and foreign policies of European nations and North America without being hampered by national loyalties. During World War I, for example, one Warburg brother headed the state bank of Kaiser's Germany, while another Warburg brother headed the U.S. Federal Reserve System.[15]

In the realm of what might be called "philosophical economics" rather than "practical economics," another Jew, Karl Marx, propounded the modern idea of "Communism" and thus played an influential role in *the* confrontation of the 20th century.

In the realm of science, Albert Einstein's special theory of relativity (1905) and general theory of relativity (c. 1916) together made possible the conceptualization of atomic bombs, and plunged us all into the daily 20th century reality of nuclear fear. It is interesting that Einstein, although he undoubtedly knew that his theory allowed for the possibility of developing a fission bomb, repeatedly denied that this was possible until Hitler began to exterminate Jews in Germany. Becoming aware of this in 1939, Einstein then wrote a letter to Roosevelt confirming that an atomic bomb seemed feasible according to his calculations and the famous formula $E=mc^2$.[16] Roosevelt then launched the Manhattan Project, ushering in the nuclear age. Even if we do not perish in a nuclear holocaust of falling bombs, our environment is already threatened by radioactive contamination from weapons testing, reactor waste and pollutants released in "impossible" (so we were assured) reactor accidents at Seascale (England), Chernobyl (U.S.S.R.) and

Three Mile Island (U.S.), not to mention smaller "events" in Canada, France and other places that have not been widely publicized.[17]

Einstein cannot be reasonably held responsible for creating the nuclear age because there were many brilliant (non-Jewish) minds working along the same lines; the secret of explosive atomic fission would have been discovered sooner or later *anyway* and probably within a few years. Indeed, Enrico Fermi and Neils Bohr knew that a fission bomb was possible, just as Einstein did, in 1939. But it is interesting that only when Hitler *began to threaten Jews* did Einstein add his cachet and prestige to the project to develop a workable bomb. The stated reason was fear that Nazi Germany might develop an atomic bomb first and thus win the war that had already begun in Europe. (Germany invaded Poland on September 1, 1939. America did not enter until two long years later, December 7, 1941, allowing Britain, Russia, Canada and Australia-New Zealand to bear the brunt of the Nazi onslaught.) In retrospect, it seems unlikely, if not truly impossible, that Germany could ever have developed even one atomic bomb during World War II. The reason was not expertise, which the Germans had in abundance, but electricity. The Manhattan Project demonstrated that to obtain fissionable Uranium 235 or plutonium in sufficient quantity to achieve critical mass for a bomb, truly immense amounts of electrical power were required to separate the fissionable elements from the non-fissionable ores.[18]

The British, who had done much valuable preliminary work on the development of a practical atomic bomb (handed over to the Manhattan Project),[19] had long since calculated that their own hydroelectric resources were totally insufficient to produce enough fissionable material to make an atom bomb. *They also knew that Germany's hydroelectric resources were likewise insufficient to perform such a task.*[20]

Only America's Tennessee Valley Authority (TVA) could do it. Only the TVA had enough hydroelectric surplus. In Eurasia, only huge Russian rivers like the Dnieper and the Volga could provide sufficient electrical power to make material for an atom bomb (which they did),[21] and this explains the Anglo-American aid to Soviet Russia during 1942-1943 as a desperate measure to keep the Nazis from defeating the Red Army and taking control of these rivers.

But the fact remains that in 1939 it was known that Germany could not produce an atom bomb within its own resources. And the fact remains that when the United States did finally make workable atom bombs in 1945, they were dropped *not* on Nazi Germany but on Japan: the weapons were "tested," as it were, on a non-Caucasoid and certainly non-Jewish population when, in the first place, the "race to the atom bomb" had begun to stop Nazi Germany from getting it first (however unlikely) and only because Hitler had begun his campaign of extermination against Jews.

Our history books tell us that the A-bombs were used against Hiroshima and Nagasaki because of the Japanese tradition of "inhuman resistance" and "fighting to the death," a propaganda theme promoted by Hollywood. But, as Ladislas Farrago relates in *The Game of Foxes*, his work with the wartime OSS revealed that Japanese/Oriental conflicts have been resolved by negotiated treaties or surrenders much more often than European conflicts; therefore, a negotiated end to the "Pacific War" could reasonably have been expected.[22] Nonetheless, and despite the fact that the OSS report had long been in Pentagon possession, the first (and only) atom bombs to be dropped in war were dropped on non-Caucasoid populations.

We are faced with the following *facts*: the current nuclear age was initiated with the help of a Jewish scientist, Albert Einstein, but only when Jews in Germany were threatened. Then, once developed, nuclear weapons were tested on a non-Caucasian population outside of Europe (where no Caucasians or Jews could be threatened).

Aside from scientists like Albert Einstein, Western Civilization has been enhanced by musicians such as Arthur Rubenstein and Yehudi Menuhin, not to mention more "popular" musicians and composers such as Leonard Bernstein. Western visual art has been enriched by Marc Chagall in painting and by Jacques Lipchitz in sculpture. Solomon R. Guggenheim established a world-famous museum and repository for modern, non-representational art.

In the area of medicine, there have been so many Jewish contributors that it would be useless and unfair to mention just one or several. However, Sigmund Freud should be mentioned because he almost single-handedly introduced the concept of mental illness that could be treated by application of his psychoanalytic method. Freud, and his numerous, predominantly Jewish practitioners, have collectively established our conceptions of mental health, proper emotional adjustment and acceptance of credible therapy throughout the Western World. Thus, Sigmund Freud and his colleagues have actually been instrumental in molding our conception of the "well adjusted" individual in Western civilization, and purport to possess the therapy needed to adjust anyone out of balance.

On a slightly less esoteric level than psychoanalysis, it is a commonplace observation that Jewish professors are inordinately represented in subjects such as sociology and anthropology—subjects of study that deal with human relations and development. Jewish experts in these fields are, again, inordinately represented on boards of education, influencing what is taught from preschool through graduate study.

In an event of paramount importance in terms of social conditioning, we find inordinate Jewish representation in, influence of, and control over mass media—whether it be Hollywood's films and television shows or newspapers, books and magazines. Television/film production and distribution, no less than mass market publishing, is an expensive proposition. It is no accident, then, that Jewish creative people on one hand, and Jewish financial interests on the other, have joined forces to create, and largely dominate, Western entertainment and information industries. If anyone doubts this, it is only necessary to observe the credits of television shows and Hollywood films. One will find a high proportion of Jewish names, with no way of knowing how many more Jews with changed names may have been involved in the production. But "behind the scenes," as it were, Jewish control is even more pervasive because virtually all of film/television financing and distribution is a Jewish specialty. The evolution of this situation, from the days of Sam Goldwyn to the present, is detailed in Neal Gabler's *An Empire of Their Own: How the Jews Invented Hollywood.*[23]

Jewish control of North American mass media gives, and has given, ample scope for subtle and not-so-subtle manipulation of public opinion, distortion of self-perception and even propaganda which has molded foreign policy. It has also permitted a gradual and incessant denigration of non-Jewish society while, at the same time dignifying and elevating Jewish culture. Films and prime time television are filled with violence, drugs and insensitive, casual sexuality—and the perpetrators of these crimes and degradation are always Blacks, "Mafia" or other non-Semites. Who has ever seen a Jewish cocaine dealer or murderer on prime time television or in a major Hollywood film? There is *one* recent exception: the film *Bugsy*, which highlights the career of Bugsy Siegal. It has been criticized as an example of anti-Semitism because Bugsy happened to have been Jewish.[24] In the movies and on TV, while mayhem is committed by non-Jews, Jews are presented as quiet, gentle, innocent victims of rampant crime.

If anyone tries to speak out against this blatantly propagandistic stereotyping, the complete arsenal of interwoven Jewish power immediately swings into action to destroy the dissident voice, a brutal response camouflaged with loud lamentations of "anti-Semitism."

In July 1991, Dr. Leonard Jeffries, Chairman of the Department of Africana Studies at City College, New York, made the "mistake" of complaining about the obvious: he was tired of Jewish-controlled Hollywood and its stereotyping of Blacks from the days of *Gone With The Wind* (when Blacks were portrayed as step-and-fetch-its) to today (when Blacks are characterized as thieves and drug dealers).[25] Predictably, and right on cue, Jewish columnists in papers like the *New York Times* wrote columns about the "ugly spectre" of anti-Semitism raising its head again.[26] But the *real* problem was that Jeffries had clashed with Arthur Schlesinger, Jr. and Diane Ravitch of the New York State

Commision on Education over revising the official curriculum to correct distorted facts about African history and to include some unpleasant facts about Jews and the slave trade.[27] The "Establishment" decided that Jeffries had to be silenced. Aside from the attacks in the Jewish-controlled press, trustees of the College were strong-armed into placing Jeffries on probation. This was on October 28. Jeffries was not quite " silenced," even though we may be sure that financial pressure was placed on the College, through the city administration of New York, by Jewish financial interests. Then, further allegations surfaced that Dr. Jeffries had threatened the life of a Harvard student journalist who conducted an interview with the beleaguered professor. Jeffries dismissed these accusations as part of "a media lynching," but they were serious enough to warrant "an investigation" by the College and possible dismissal of Dr. Jeffries due to "conduct unbecoming a member of the staff."[28] Rest assured, every effort will be made by the Jewish mafia to fire Dr. Jeffries—even if the reasons have to be invented.

A similar thing happened at Vanderbilt University. On September 20-21, 1991, the author was asked by a civil liberties group (supported by Jewish businessmen in the Nashville area, in conjunction with Vanderbilt University) to talk about *The Iceman Inheritance*. The end of the talk branched off into the theme of this book: that if Caucasians as a whole were more aggressive than other major groups of humanity, then the Jews could be considered as an "extra-aggressive" cultural and ethnic expression within the Caucasian race, and this accounted for their inordinate influence in Western civilization. The lecture also pointed out that the vast majority of contemporary Jews have no historical or genetic relationship to ancient Palestine at all; most of them are descendants of Russian steppe tribesmen who were converted to Judaism en masse around 740 A.D. Being of highly aggressive Caucasoid genetic stock, modern Jews could be expected to be highly aggressive in consequence.

Now, the Khazar-conversion origin of most of modern Jewry is a fact discovered *by Jewish scholars* and a fact accepted by almost all Jewish scholars. The contention that these Khazar steppe tribesmen would have been highly aggressive is simply an extension of the central argument in *The Iceman Inheritance.* Aside from mere theory, the Khazars' own history shows them to have been highly aggressive and successful warriors.

But it was fact that contradicted Jewish myth, and which undermined Jewish self-conceptions as being gentle victims of anti-Semitism. It was fact that years of Jewish propaganda had sought to disguise, distort and muddle. Justification for the creation of Israel "as a Jewish homeland" depended upon the purposeful distortion of the truth among the propagandized mass of North American voters. A clear exposition of the truth was, therefore, intolerable to the Jewish financial interests who backed the civil liberties group. By November 15, 1991, a Jewish lawyer acting for Nashville's Jewish financial interests, which had backed the lecture, threatened to sue![29] Have things gotten to the point that a person can be sued for telling the truth about an historical episode that happened 1300 years ago? Apparently, if that episode is something which the Jewish media-financial-educational-legal complex wants to keep suppressed.

Jews represent less than five percent of the population

of any Western European or North American country.[30] Yet, as the summary of their influence over the past few pages illustrates, they have somehow come to influence Western civilization to a degree out of all proportion to their small numbers.

What is the explanation for this? Are they really the chosen people, somehow having some divine blessing or power that makes their prominence inevitable?

This book proposes an explanation for the anomalous role of Jews in Western civilization. Like *The Iceman Inheritance*, it is a rather mundane explanation relying upon historical and anthropological facts. There is nothing particularly arguable or controversial about it.

Nonetheless, this explanation, by presenting the facts, will destroy the mystique of Judaism for Jews and gentiles alike. It is long past the time when the Jewish mystique should have been subjected to objective scientific scrutiny. The world can do without the mystique, the myths and the delusions, which have caused a great deal of human tragedy.

In order to grasp this great Hebrew hoax, fundamental ideas presented in *The Cronos Complex* and *The Iceman Inheritance* must be restated. The ideas are important enough that they must be expounded in adequate detail.

Animal Territoriality

A true revolution in biology began, modestly enough, with a group of English bird watchers around 1920. One can imagine Edwardian ladies and gentlemen gently intruding on the predawn of their Somerset water meadows, chilly in February, fired with their eccentric enthusiasms. Armed with an assortment of field and opera glasses and picnic baskets full of goodies from Fortnum and Mason—one imagines—they awaited a momentous event. This was the return to England of a certain species of warbler, a species which had migrated southward to Portugal and North Africa, in order to wait out the mild English winter. They always returned about the same time, mid-February, and the English regarded their singing as a sure sign of spring. In that time, it was still possible to believe that songbirds sang for the pure joy of life. And so the arrival of the twittering flocks of warblers was an occasion for springtime celebration among these Somerset ladies and gentlemen. Being dedicated bird watchers, they might wait for two or three crisp February mornings before the Great Event happened, but when it did everyone was delighted and celebrated with toasts of brandy (for the gentlemen) and sherry (for the ladies) from those picnic baskets. Then they made notations about the date and time the first flock had arrived, the number of birds in that first returning flock, and so on in a most meticulous manner; and then they went home—or so they would have us believe.

One member of this genteel group, Henry Eliot Howard,[31] often stayed behind after his companions had left. He observed the birds more closely and scribbled some interesting notations in his notebook. First, the initial flocks of birds that arrived in mid-February were composed of males only. Female warblers usually did not arrive for another two weeks, but no one waited for them with the same dedication. Then, Howard noticed that when the males arrived they would fight and flutter among themselves, apparently to gain possession of some perch among the cat-tails and shrubs along the water. The male birds competed quite seriously to obtain a particular perch and, if and when they were successful at doing so, they would sing their hearts out. Over the course of several years of careful observation, Howard thought that his particular flock of early-returning male warblers seemed to have preferred perches that provoked the keenest struggles. Eventually, however, all of the male warblers found a perch, although it might not have been the one of their dreams, and commenced to sing. Strangely, Howard noticed that while the birds twittered in flight, or when competing for a perch, they only sang their customary melodious warbler song when actually sitting on their hard-won perches.

About two weeks later, as usual, flocks of female warblers would arrive and they, too, would twitter and seem to jostle and fight for a few days. Eventually, Howard realized that these females were competing for the males who had the best perches and who sang the loudest. The most aggressive females, then, fought to be with the most aggressive males. Within a few days, all the warblers had sorted themselves into couples, and began to build nests around the first week in March.

Howard's gentle hobby of bird-watching revealed a fact of truly fundamental importance about animal behavior, a fact with far-reaching implications. The warblers did not sing for the pure joy of life, as a more romantic generation of Victorian naturalists had been content to assume: they sang in order to assert territory, their little perches. The male warblers competed for preferred territories, and the female warblers competed to mate with the males who had succeeded in claiming and holding a desirable perch.

The 1920's was blessed with a plethora of "naturalists"; mostly these people were enthusiastic amateurs, but many of them had extraordinary observational skills and after Howard's observations were published in an obscure, but avidly read, local Somerset naturalist magazine, other bird watchers confirmed the same phenomenon with other species. It seemed that *all* songbirds claimed and asserted territory and acquired nesting partners, exactly like Howard's warblers. Within months, all sorts of naturalists, not just bird-watchers, took up the research. A male robin—not a perching songbird—claimed about half an acre of land, would sing to assert his ownership and would chase all other male robins away. Then female robins would compete to mate with the male robins who owned the largest real estate.

A squirrel will claim two or three trees in the same way and with the same reproductive result. A bull seal on the Pribiloff Islands will claim and defend an area of beach before the female seals arrive; a stickleback fish will claim a square yard of stream bed and chase all other male fish away; Uganda kob antelope are a little stranger because, within this species, the males fight to acquire one of a very limited number of clearings in the bush ("arenas"), and most males end up without an arena.

Very quickly it was suggested that the territory had economic importance to the animal; that it contained a sufficient food supply for, initially, the male, and later his mate. But this explanation just as quickly crumbled under the evidence: a songbird's perch had nothing to do with food supply; it was a perch and nothing more. Birds would fly over a wide area to obtain their food but would always return to claim the perch. The nest might be located on the perch, or near to it, but it might not. As for the bull seal's patch of beach, it contained no food whatsoever. A pair of squirrels seemed to "own" two or three trees, but they obtained nuts and acorns from many. As for a kob antelope, the grassy clearing in the midst of the bush did not contain enough grazing for one animal and, in any case, no male kob antelope has ever been seen eating while occupying his hard-won arena.

Territoriality seemed strange and inexplicable animal behavior. Over many years, however, the importance of territory became apparent to researchers. It was of no economic importance (in most cases), but was of crucial reproductive importance. Seals, possibly, provide the best example. Bull seals arrive on the breeding-beaches-to-be before the females, and the bulls proceed to fight to obtain a part of this rather limited stretch of beach. The largest, most powerful, courageous and confident bull seals naturally win the largest tracts of beach; bulls of lesser prowess must be content with smaller real estate, and the smallest, weakest and least confident bull seals cannot win a piece of the beach at all. When the female seals arrive after this has been sorted out, not only does a large bull seal territory seem to attract them, but a large territory can accommodate more of them. These females become the bull seal's harem and he breeds with them. It is obvious, then, that the largest and "best" bull seals will pass on more of their successful genes to offspring.

Kob antelope of Uganda exhibit a very similar behavior. There are only a few arenas, and so only a few top bulls can occupy one. Female kob antelope will consent to breed *only while standing on an arena.* Thus, only the few exceptional kob bulls impregnate the entire herd, and most male kob antelope never breed at all. Kob territoriality insures that the best male genes (at least) are represented in all resulting kob offspring.

Most species are not so extreme as seals and kob antelope. With Howard's warblers and most other creatures, most animals breed—but, the "best" breed with the "best," the "average" breed with the "average," and the poorest examples breed with the poorest of the species. For example, in any given warbler flock there may be one or more hapless males who can obtain nothing but the poorest and least desirable perches. If there are enough females, it is possible that a very noncompetitive female will end up breeding with one of these males—but it may be that there are not enough females to go around, and so one or more males will not breed. And they will not sing.

Gradually, territory was revealed to be *a behavioral mechanism permitting the assertion of biological relevance.*

Animal Dominance

Meanwhile, other researchers were investigating an equally curious phenomenon which they called "animal dominance." Some species did not seem to care for real estate at all; but they did seem obsessed with status within a group. The observations of Konrad Lorenz have become classics in this field, eventually winning him a Nobel Prize, and it will be more than sufficient for our purpose to present his research.[32]

Lorenz did much of his work with jackdaws, a kind of European crow. Like all crows, jackdaws are both intelligent and gregarious. They travel in large flocks; a flock may have a home base territorial area, but a flock will range widely for food even if its nests are concentrated in a rather small domain. Individual birds are not territorial. They are, however, status-conscious. From almost the instant of his hatching, a male jackdaw will begin to compete with other males for "dominance"—which is to say that he will try to intimidate other males and force them to defer to him. He will begin pecking at other males, begin to threaten them, and discover whether an adversary backs off (defers) or takes up the challenge with enthusiasm so that *he* must back off. Jackdaw males do this throughout their lives in a never-ending competition. Any given flock is arranged in a hierarchy of social status. There is a number one male to which all other males (and females) must defer: he gets the first chance at food, the choice roosting place, and leads the flock. There is a number two male to whom all birds (except number one) must defer—and so on down the social structure to that hapless male at the bottom of the pecking order who must defer to every other male in the flock.

The female jackdaws compete among themselves in exactly the same manner, but with a specific goal in mind: they compete in order to breed with the highest ranking male that they can possibly get. So, a jackdaw flock consists of paired couples: a king and queen, various ranks of nobles, various ranks of commoners below them and wretched families of serfs at the bottom of the pecking order. The best jackdaws breed with the best and, because higher social rank ensures preferred access to the necessities of life (like food, better roosts and sheltered nesting sites), the best jackdaw genes are much more likely to survive in a greater number of offspring.

Robert Ardrey, who popularized these new biological concepts in the 1960's and early 1970's in books like *African Genesis* (1961), *The Territorial Imperative* (1966) and *The Social Contract* (1970), continually maintained a separation between territory and dominance. Perhaps this was only because researchers in the two areas perpetuated a different jargon themselves, and Ardrey felt too unsure to challenge them. Ardrey always seemed haunted and disturbed by "dominance," and wrote:

> There are ghosts in evolution's attic. A shutter slams on a windless night. There is a scraping sound above our heads... then there is silence. We rush out the door. There is no one. Who was it? Where did he go?[33]

The situation, in reality, is not haunting or disturbing. A social position, in a jackdaw flock, has a certain "size"—the prerogatives attached to it. It is no more or less than "social territory," and when asserted, has exactly the same reproductive relevance.

In *The Cronos Complex* (1973), I wrote:

> I should warn readers here that most naturalists still separate "territory" and "dominance." It is uniquely my quirk, so far as I know, to equate social position and spatial territory as being functionally similar: mechanisms for manifesting biological relevance. I therefore rather loosely use the term *dominance* to mean: "the comparative success of an individual in manifesting biological relevance according to the mechanisms operative in his species." I reject the contemporary idea of "dominance" which refers specifically to social position. In my terms, dominance may be manifested through possession of territory *or* social position as they are a function of biological relevance.[34]

Double Assertion

Some animal species (kob antelope, seals, robins) enter into competition for actual territory, while other species (jackdaws, barnyard hens) concern themselves with struggles to attain social position—and some species do both at the same time. It may be that the more intelligent a species is, the more mechanisms it requires to assert biological relevance. The so-called "highest animals," the primates (monkeys, apes and men), apparently need to compete within their groups for social position while, at the same time, groups struggle with each other for control of geographical territory.

Field observers such as C.R. Carpenter, George Schaller, Jane Goodall, Eugene Marais and Dian Fossey all studied gorillas, chimpanzees or howler and rhesus monkeys at close range, and for years on end collected a wealth of observations that prove that, within their own groups or closely-related family clans, those primates behave very similarly to jackdaws: they strive to attain the highest possible social position. As in most of the animal world, the competition is usually bloodless—at least, more bloodless than is the case within comparable human struggles for social position. Generally, threats and gestures suffice, and apes rarely come to blows.

But not always.

Gorillas and chimps are not always so peaceful. There appears to be some motivation in them that goes beyond mere striving for social position and which hints at some knowledge of past and future—a dim, disturbing knowledge that provokes tragic behavior. Dian Fossey observed that sometimes when one male gorilla had usurped another, usually older male's leadership position, the new dominant male would murder his deposed rival's offspring.[35] It seemed as though the new dominant male was compelled to erase all evidence of his predecessor's very existence. (This is a most important observation, full of tragic implications for humanity, to which we will return after a very brief discussion of group territoriality.)

In addition to the continual tension of individuals within a primate group or family clan competing for social position, most primates compete as a group, against neighboring groups, for the possession of actual geographical territory. Again, among monkeys and apes this is a relatively bloodless activity compared to human warfare. Threats and gestures suffice, aggressive behavior that keeps each group on its side of the boundary; seldom do matters come to blows. Some monkeys have even developed special adaptations for this sort of territorial warfare. Howler monkeys of Central and South America yell; they have special cartilagenous resonance chambers in their throats that can produce an incredible volume of sound. In Howler monkey warfare, two opposing groups push each other back and forth across the boundary by sheer force of decibels.

Because territorial conflicts are relatively bloodless in the primate world, it does not mean that things are static. Sometimes, one group will take over the territory of another group, and absorb the conquered group, by sheer intimidation. Losers are at first reduced to the lowest status within this new and larger group, but gradually there are trials of dominance, and rebellions, until there is a new social order within the new group—one that eventually recognizes the relative dominance of all the individuals comprising it.

Individuals can make a difference in the primate world, and a case can be made for "the great man" theory of history. In one experiment, administered and recorded by C.R. Carpenter, seven separate groups of rhesus monkeys were released onto an island off the Georgia coast. The experiment was undertaken simply to observe rhesus groups in action, and it was confidently assumed that the original seven groups would maintain a relative stability. And they did, for several years. Then, a monkey of remarkable charisma was born to one rhesus group. Before he became an adult, he was dominant within his own clan; and then he rapidly began a program of conquest previously unrecorded in the annals of animal observation. He conquered all six other groups on the island and welded them into a rhesus empire with himself as emperor. Such a chain of events had never been observed before in the wild, and it must be very rare. Naturally, Carpenter called this monkey "Napoleon."[36]

In the operations of animal territoriality and dominance, Robert Ardrey saw the reflection of human culture. We, too, compete within our groups for the highest possible social position, and we, too, indulge in group warfare with rival groups for the possession of geographical territory. For Ardrey, there was no essential difference between animal behavior and human behavior; a few hundred thousand years of evolution could easily account for the material differences between ourselves and, say, baboons, given our greater intelligence. And this material richness, in turn, explained the greater complexity of our struggles for social dominance and territory. The only problem, for Ardrey, was that our gradually acquired material richness had deluded us into thinking that we were "above" the biological imperatives of elitism. Ardrey believed that we were diluting the necessary struggles for territory and social position through the invention and introduction of various "social safety nets," of which socialism and communism were the worst. He saw the progressive degeneration of humanity as we nullified the basic biological realties. In *The*

Social Contract, Ardrey argued passionately and eloquently for a return to some principles of biological elitism—dominance asserted through aggression—before we degenerated ourselves out of existence.

Although, in matters of practical social policy the author has a certain sympathy for Robert Ardrey's perspective, Ardrey was simply *wrong* in equating the animal world with the human one. Primate territorial behavior and social dominance behavior, for all its similarity to human antics, *does not* completely and adequately mirror human culture.

The Cronos Complex

There is one characteristically human culture trait which is not mirrored *at all* in any other known animal species. That culture trait is *religion.*

In this day and age many of us take a cynical view of "religion," and would consider it to be a more or less irrelevant invention of the human intellect, a kind of pretentious delusion invented only for fooling ourselves into the comfort of thinking that we biological creatures have some claim to a "soul" or "spirit" that makes us special. Karl Marx dismissed it as "the opiate of the masses," a comforting delusion that drugged people into caring too little for their material welfare, an opiate prescribed by those who would oppress the average man and woman, a kind of anesthetic to dull the hurt and pain of exploitative socio-economic manipulations. Religion has performed all of these roles.

However, from a biological point of view, *religion is the conceptualization of time before birth and after death.* The conceptualization of time may be expressed allegorically, that is, by invention of, and belief in, various supernatural beings or powers. Gods, spirits, angels, etc., all have one thing in common: they are "immortal," or at least exist much longer than any human being, and thus exist in an extended past, present and future that is much larger than any individual human life. Even among one of the earliest human religions yet studied, the beliefs of the *orang pendek* of modern Indonesia, there's a vague sort of ancestor reverence (past orientation) and then belief in a "Grandmother Longpaps"[37] who will wash sinful souls in boiling water after death (future orientation); these beliefs assert the conception of time before an individual's birth and after an individual's death in which, somehow, humans still "live" or "exist" and with which the living have some sort of connection.

We have absolutely no observational data suggesting that any other (known) kind of animal species on earth has made such a conceptualization: we have observed no priests among seals, no shrines among robins, no rites among jackdaws that suggest any sort of "life-transcending" idea of time. This seems to be a purely human conception; and such an abstract thought may be possible only when the brain has reached a certain size, or has become complex enough to make possible certain neural connections. There are hints that perhaps some of the more intelligent apes, gorillas and chimpanzees, may be just below this threshold of neural complexity and that some may approach it.[38] Dian Fossey's field observations, documented by *National Geographic* photos,[39] reveal that a newly dominant male gorilla murdered his deposed rival's child. This was a shocking revelation to the *National Geographic* team, and Fossey had never witnessed such behavior before (nor had George Schaller before her). This sort of behavior must be very rare among gorillas and suggests something quite profound: the newly-dominant male gorilla must have been an exceptionally intelligent one; his brain was almost sophisticated enough to grasp the idea of life-transcending time. It was not enough for him to kill his rival to become dominant in his group; it was somehow also necessary to kill his rival's *extension* (offspring) who, somehow, *was* his rival *still,* in a way that might threaten his future dominance. Now, the interesting thing is that all dominant adult male gorillas seem to regard younger males with a fair degree of suspicion, knowing full well that the younger ones will one day challenge them. Yet, even so, the older established males do not try to kill off these future rivals before they can mature.

In the case of this murder, however, the new dominant male fused the identity of his deposed rival with that of the rival's child. *Both* were somehow the new male's rival in a dim conception transcending death (the adult rival's death). There must have been the notion that *identity* (of the dead rival) could exist beyond his physical death and continue to be a threat to the newly-dominant male. This is almost a human-like conception of time, one might say a "flash" of realization, but it did not last and had no obvious behavioral repercussions. After this passion of murder, the newly-dominant male acted like a normal gorilla: he did not fashion any crude representations of his rival, or the dead baby, and symbolically "re-kill" them at intervals. Neither was he overly aggressive toward other young males growing up within the clan.

We have introduced a new word, "identity," and have seen that even a gorilla can conceive of it and kill because of it. We have seen that the possession of territory or social position has biological relevance—but does it have some other relevance we have not yet grasped? In 1949, a researcher named J.C. Braddock conducted an interesting experiment on platys, a common tropical fish often used by animal behaviorists. Ardrey recounted this experiment in *The Territorial Imperative*,[40] stating that Braddock put each of the fish in a small aquarium and put a little duckweed in half of the aquariums. The platys boasting duckweed he termed "residents," those without duckweed he called "intruders." After giving all the fishes a few days to get used to their homes, he put the "intruders" into the "residents'" aquariums. Although some of the intruders were much larger than some of the residents, in nearly every case, the resident fish maintained dominance as long as they were together.

In another similar experiment, but one involving crickets,[41] a number of these insects were provided with little boxes, "niches," as homes where they could sing: a territory. A cricket will not sing unless it is in possession of territory. Crickets are fighters, and so the experiment involved provoking crickets to fight: this was done by randomly taking crickets from their own boxes and placing them close to the boxes of other crickets. All the crickets were involved, at one point or another, as both "intruders" and "residents." The researchers discovered that, *without exception*, the winning cricket was the one nearest to its niche (something that Chinese cricket-fighting gamblers have known for three thousand years). Some of the crickets were, of course, larger and stronger than others, but it was proximity to territory that counted.

This might seem to contradict previous examples of territorial behavior, but it does not. In the competition to acquire "virgin territories" (as with flocks of returning warblers or robins), it is size and strength that usually decides the outcome. However, once territories are established, the defender has an advantage that can, to a remarkable degree, compensate for size and strength differences.

In the example of the murdering gorilla, we saw that this thing called identity, though necessary for reproduction, can also provoke lethal violence. In an experiment performed by J.S. Brown in 1961, it is apparent that this thing called identity, while necessary for successful reproduction and the manifestation of biological relevance, is something *more*. It is something so precious that, to preserve it, an individual will go without the opportunity to reproduce and will die defending it.

Using platys again, J.S. Brown put several fish into a large aquarium and waited for some time until the fish had sorted out their "pecking order." Platys, like jackdaws, are not strictly speaking territorial, but compete for social status. When Brown had observed the platys sufficiently to determine the hierarchy they had established, he began slowly to lower the temperature of the water. The water became so chilly that the fish lost all interest in eating; a bit colder, and they lost all interest in sex. The temperature was lowered still further so that death was inevitable and, all the while, the platys *continued to compete for dominance*. They died defending their social positions; *that* social position was the bastion of individual identity for platys and they preserved it to the last spark of life.[42]

So, it seems that "territory" (whether actual real estate or social position) is not only a "mechanism to manifest biological relevance," which it most definitely is since most species cannot reproduce without it (or it crucially affects the quality of offspring)—territory confers something even more precious to the individual: "identity" or "confidence," and the individual may do violence to preserve it or may prefer suicide to relinquishing it. This is an observation of extreme importance to purely human behavior.

If religion is the human conception of time before birth, which *has been* "inhabited" by ancestors, and after death, which *will be* "inhabited" by future generations, then time becomes a kind of territory. It is a sort of intellectual and abstract expanse of forest that can be inhabited, not wholly different from a vast human clan, or group, in which the "pecking order" of dominance involves the dead, the now living and the yet-to-be-born.

In short, once humanity conceived of time before birth and after death of an individual, time was immediately and naturally viewed as a kind of territory. And as such, it was necessary to assert and defend our "habitation" of it.

In *The Cronos Complex* (1973), the author defined modern, cultural man as *Homo temporalis*,[43] a creature who has conceived the abstraction of time and perceives it *as a racial environment.* This seemed more apt to the author than *Homo sapiens*—"reasoning man"—because many experiments have proved that many animals can "reason" out problems to an astonishing degree, and we humans have no monopoly on logical thinking. Homo Sapien is not a truly distinctive definition of what it means to be a "man." We became "man," as distinct from just another intelligent animal, when we conceived of *time* as *a racial environment.* This forced us immediately to begin to invent *human culture* which differs in quality from anything developed in the animal world.

The English anthropologist, Edward Taylor, has offered a definition of human culture that has been widely accepted: "that complex whole which includes knowledge, belief, morals, law, custom and other capabilities acquired by man as a member of society."[44] But this is not really a definition of what human culture *is*—it is merely a list of symptoms that are manifested when culture is acquired and practiced. Human culture *is non-genetic communication across the barriers of birth and death permitting human habitation of time as a racial environment.* That is a truly biological definition of culture, one that evolved logically from other known forms of "territorial" and "dominant" behaviors. It is the definition the author presented in *The Cronos Complex* in 1973 and expanded in *The Iceman Inheritance*.[45]

The problem with time as territory is that it, like any other territory or social position, must be asserted and defended if it is to confer identity. This means that each human individual must assert his/her lifetime *against* the stature of the past, and *against* the threat of the future, if he/she is to acquire identity in a purely biological sense permitting reproduction. Our past and future become territorial rivals. A simple example can illustrate the entire human dilemma.

No other territorial animal would think of competing with a rival who is already dead! Territorial competition among the living is the only relevancy in the animal world, and (as we have seen) it confers identity and the confidence to affect reproduction. But a human writer may well compete with someone who is already dead; a writer may strive for the honor of writing a better novel than Melville's *Moby Dick*, a better play than Shakespeare's *Hamlet*. An artist might strive to paint a more exquisite portrait than Leonardo's *Mona Lisa*; a sculptor might struggle to surpass

Michelangelo's *David*; a musician might strive to be judged better than Mozart, Bach or Beethoven.

This sort of "non-genetic" territorial struggle involves *creations* which are a valuable part of our culture—but it is easy to see that the people who engage in this sort of territorial struggle are usually not very successful in purely "biological" and practical terms: most writers, artists and musicians are poor; many have died in abject poverty without children or without the means to care for them.

The problem with time as an environment and "territory" is that it is non-biological. Those who can conceive it most clearly, and who wish to assert their identity within the past-present-future continuum, *tend to be biological failures.* Priests, nuns, great writers, artists, painters and scientists—all tend to be statistically sterile. In order to be successful, biologically and practically, one must largely "shut out" deep considerations of the human past and future, one must almost refuse to create "culture."

We see now that Robert Ardrey was very wrong. Human culture, as distinct from purely animal culture, has been developed by *biological failures.* If we had preserved the "imperative" (as Ardrey would have it) of purely biological elitism, as it is faithfully preserved in the purely animal world among all other species, *we would not have developed our complex human culture and we would not have progressed significantly beyond the ape stage of technology.*

Here's why.

Once humanity conceived of this thing called "time" as a territory to be "inhabited" it was probably not well understood by most people. But it was necessary for us to open lines of communication with our ancestors (at least, and at first) if we were to assert this strange kind of territory. Some individuals among early human groups understood the abstraction a bit better than their fellows, or claimed to understand it better, and claimed to be able to communicate with the ancestors inhabiting the past, the dead. These people may have been crazy, or they may have been exceptionally intelligent (or both), but being of exceptional intelligence, in at least some cases, they could grasp cause-and-effect relationships better. They became our first priests, shamans. Even though a shaman may not have been a great hunter himself, he was supported by the group because of his supposed ability to communicate *through time* with the ancestors.

Now, the shaman, although he might not have been a good physical specimen, nonetheless, was a bit more intelligent and grasped cause-and-effect. He could figure out, for instance, that it would be easier to approach a herd of game animals if the hunters would disguise themselves in animal skins. The first known representation of a human shaman is called "The Great Sorcerer" and is a cave-painting from France; it shows a human being inside a deer-skin, wearing antlers. He seems to be acting out deer-like movements. Perhaps he was instructing hunters, in a ritual dance before the hunt, on how to *become* deer. Done successfully, the hunters should have been able to approach the deer herd more closely and to have killed more of the animals. The shaman's magic of "shape-changing," still a vital part of some primitive Caucasian, African and Mongoloid cultures, directly resulted in greater hunting efficiency.

Again, in purely biological terms, what has happened is this: *the human conception of time as a territory com-*

pelled us to preserve exceptional intelligence that was of no use in the "animal present" but which turned out to be of future benefit.[46] Our conception of time as a kind of territory began the process whereby humanity started to preserve exceptional intelligence by establishing cultural havens for extraordinary people who, likely as not, would have perished in the purely practical, biological world which has but one command: "Survive! *Now!*"

Once the preservation of exceptional intelligence begins, "progress" tends to become a geometric function. Our first haven from the "biological present" was religious: we supported shamans because they could communicate with the past, with ancestors. But their intelligence and "theoretical" approach to problems resulted in greater group efficiency (as illustrated by the advantages of shape-changes in hunting) and this extra margin above the demands of mere survival meant that human groups could begin to preserve yet other exceptional people. Eventually, we preserved colleges of shamans, priests and priestesses, who are thought to have invented writing. With shamans, we had opened up communication with the past, our ancestors; with writing we opened up *non-genetic communication with the future,*[48] the generations yet-to-be-born. It is very possible that priests and priestesses conceived the idea of agriculture and animal domestication; the growing of food against future need. Certainly, the oldest known religions are fertility cults associated with the growing of grains, but with the fertility goddesses often having attributes of domestic animals.

When did this purely human conception of time as a territory begin? It is impossible to state with any certainty. Indications of such a conception would be burials in which some effort had been made to supply the dead with provisions for their "afterlife." That would be evidence of "time-factored" living. The oldest deliberate burial, with grave goods, is presently a Neanderthal interment in France. It is about 50,000 years old. From about the same epoch, Marshack found bones scratched with notations of lunar phases showing that humanity had begun to count time.[49]

More research has been done in Europe than anywhere else, and it is probably only for that reason that the first deliberate interment has been discovered in France. Humans probably first conceived of time as territory far earlier than 50,000 years ago, and it could have happened anywhere (Africa is probably the best candidate)—but the evidence has not yet been discovered because archeology outside of Europe is still a rather sketchy affair.

It stands to reason that if time is a territory, and if our own past and future then become territorial rivals for identity, then human culture is not only a mechanism for peaceful communication with the past and future, but equally a mechanism to defend each "present lifetime" against the past and the future. *The Cronos Complex* and *The Iceman Inheritance* outline in considerable detail the cultural ways that we try to "surpass" the past (by progress of various sorts) and try to "hold the future back" (by warfare, and by social norms and values imposed on the young) so that each human generation may attempt to be unique and so preserve its identity.[49]

And here comes the psychosexual crux of humanity: if offspring are sure to be a threat to our identities, then why would we bring that threat into being? Why would we reproduce at all?

The answer to that is nothing more sophisticated than pleasure. It has always puzzled biologists that sexual adaptations seem to become more sophisticated as intelligence increases. For example, in so far as we can estimate animal intelligence, the frequency of ovulation and the complexity of breeding behavior seem to increase as intelligence increases.

Sexual adaptations culminate in "man." As Desmond Morris in *The Naked Ape* explained for millions of titillated readers, humanity has more and more sophisticated sexual adaptations than any other known species.[50] Our sexual organs are, for example, much larger than a gorilla's even though the average gorilla is about twice the bulk of any man. These larger sex organs contain a much larger number of nerve synapses that provide a high degree of stimulation and sensory pleasure. Most of humanity is fairly hairless and this enhances the tactile stimulation derived by bodily contact. A number of anthropologists are inclined to the opinion that human female breasts, which are unique in the mammalian world in protruding all the time and not just when nursing, are a kind of "surrogate buttocks" for front-to-front copulation to remind and recall the buttocks that stimulated us when we copulated from the rear like other animals.

This explanation for the existence of human female breasts is not convincing. In the first place, apes have neither breasts nor buttocks, and these closest relatives to humanity apparently do not need the stimulatory vision of buttocks to inspire copulation. Human female breasts may have been "institutionalized" very rapidly once the conception of time as territory had been made, adding a new dimension of ambivalence and aggression to sex. Being "institutionalized" means that females with some extra chest development may have become immediately preferred by (now) super-aggressive males because their breasts could be squeezed *from a rear-mounting position* as a copulatory aggression-displacement activity: copulation was emotionally painful for the male, and he could make it physically painful for the female. As simple and as brutal as that.[51] If this is so, then flat-chested females were quickly "selected out" because they were not bred as often; breasts began to be a cultural commodity and so those females with slightly greater breasts development were gradually "selected in," by selective breeding, until breasts reached what we consider normal human proportions. This might have required only 20 to 30 generations, perhaps seven or eight hundred years, within any one human population, if our own success at producing, say, chickens with larger breasts is any indication.

As for buttocks, they came with erect stature and true bipedalism. They serve to hold the trunk of the body upright. No ape has buttocks for the simple reason that no ape is truly bipedal. All apes are quadrupeds; they walk on all fours. If some of them can stand upright, it is only because their arms are long enough to reach the ground from a standing position. Apes occasionally stand, or even walk or run on unsupported hind legs, but never for long and never for any great distance. They are not bipeds; they do not need, or have, buttocks for this reason. It cannot be denied, however, that human female buttocks are more generously developed than bipedalism alone demands.[52] Again, this may derive largely from cultural selection and subsequent selective breeding. (Grasping a female's buttocks is an aggression-displacement activity much in the same way as grasping a female's breasts.) Some females without very much chest may have had suitable buttocks, and so the two types were "selected" together and eventually blended into what we consider to be the normal human female shape. This shape may have come rather suddenly on the scene, through selective breeding, between 50,000 and 300,000 years ago. Reversing the opinions of Desmond Morris and a number of other anthropologists, it may be that the development of enhanced buttocks began to promote the custom of front-to-front copulation since buttocks are graspable, by the male, in that position.

Leaving aside the universal human development of breasts and enhanced buttocks, some races developed fairly unique sexual adaptations: the "Capoids" exhibit an elongation of the labia minora (popularly called the "Hottentot Apron") which has been observed in some Ancient Egyptian female mummies (the "tablier Egyptienne"—Egyptian apron), proving that Capoids were represented in this Nile civilization.[53] Male Capoids, from birth to death, have a semi-erect penis.[54] Mongoloids, though typically having the smallest breasts and flattest buttocks of major human genetic groups, are the most "glabrous"—the most nearly hairless (Orientals called Caucasians "hairy barbarians") and highly value the refinement of tactile stimulation.[55] Although it is disputed as a "myth," it seems to be true that Black males have proportionally larger penises, as a rule, than males of other major human groups while Black female vaginas are not so large—a situation conducive to more stimulation.[56]

All of these adaptations must have some purpose, however they evolved, and this would seem to be to overcome, by pure pleasure, any ambivalences toward reproduction that "territorial aggression" might compel. Frequent human periodicity worked to ensure that any given intercourse had a fair chance of resulting in conception. Even if these adaptations do not always work to incite the reproduction of people with exceptional intelligence, they do the job for people with average intelligence—but not all together perfectly or "humanely."

We have seen that in most species, from Howard's warblers to Fossey's murderous gorilla, the primary *territorial* urge seems operative in the male; the female's quest is to obtain the best possible male as her mate. In the male, this territorial urge conferring identity is manifested through *aggression.*

We should not expect that things would be much different when it comes to urges to manifest identity in the "territory" of time. Aggression would be the means of manifesting the "territorial imperative," and the natural object of this aggression would be the *female* who is capable of bringing the threatening future, the identity-stealing child, into being. Yet, the absolute imperative of biology is to reproduce and, however "angry" or "ambivalent" the male might be, he has the urge to reproduce in order to *assert* his identity. A fine irony and a "Catch 22!" The sexual adaptations just discussed go a long way toward dulling the realization of this irony and serve to tip the balance of emotional ambivalence toward intercourse and reproduction through intense pleasure and release of simultaneous aggression (breast and buttock grasping).

However . . .

Human sexual adaptations are not sufficient to displace or absorb all of the territorial aggression. There is a lot left over—among all human males of whatever "major genetic group." Post-coital depression is a typically male malady. All human males, of whatever "major genetic group," have institutionalized anti-female social customs that, at base, are a residue of *aggression* deriving from the "time as territory" conception. In the caves of Pilaga, in Spain, paintings show that cave-dwelling males lacerated the breasts and distended abdomens of pregnant women;[57] Orientals (of some cultures) bound the feet of their women and virtually crippled them; some Africans have institutionalized the practice of "female circumcision," more brutal than removing the male foreskin, where the major and minor labia are cut away along with the clitoris. The message is loud and clear: women will not enjoy an act that threatens male identity; female infanticide has been common human practice and menstrual taboos are a common human pathology.

Even today, in the modern Western world, "sexual abuse" of women and "child abuse" are being revealed as a major social problem previously disguised and hidden by almost total male control of social norms, values and media. Now that control has slipped, due to women's liberation and job equality, women in the media have begun to reveal that the psychosexual environment within nice, suburban homes may not be much of an improvement over Freud's brutal and "perverse" Ice Age caves.

Cronos was the Greek personification of time. English words *chronology* (time-sequence) and *chronometer* (time-measure) still derive from his name. In Greek myth, Cronos was married to Rhea ("Fertility") and so, together, they were the measure of all reality: time and fertility—creation. But the problem was that Cronos proved to be a jealous personality. He feared that any offspring would usurp his position. And so, when Rhea produced children, Cronos would swallow them. Rhea was desperate, for her joy (after all) was fertility. At last, in an effort to dupe Cronos, she offered not her latest child, but a rock wrapped in swaddling clothes for Cronos to swallow. Greedy for ultimate, eternal identity, Cronos gulped the bundle down without bothering to check it carefully. Rhea hid the child on Crete (some say in a cave, with goats to feed it). The child was Zeus, and the worst fears of Cronos were realized. Zeus survived, grew up and became powerful enough to challenge Cronos. Zeus defeated Cronos, castrated him, and threw Cronos into a deep abyss. Zeus then ruled over the Greek pantheon on high Olympus.

This seems to be a true (albeit allegorical) understanding of the human personality once "temporal territory" had been conceived. It inspired the title *The Cronos Complex* for my first book about the human psychological condition. This "cronos complex" was a craving to attain identity within the territory of time.[58] It was a malady experienced by all of humanity, with all of its triumphs and tragedies in consequence, but a tragedy softened for most of humanity by the sweetness of Rhea's tricks (sexual adaptations). In most of humanity, the aggression of Cronos was thwarted, if not perfectly, by sexual aggression-displacement mechanisms that worked "adequately" if not "humanely."

However, in one "major genetic group" of humanity (Caucasians), Rhea's tricks have been betrayed by climatic factors beyond human control.

II
Neanderthal-Caucasoids and Mountain Men

No one can say with any certainty how long "modern humanity" has existed, or where it originated. Since most of the "man-tending" creatures (and there were a lot of them) seem to have arisen in Africa, it seems reasonable to assume that "modern man" also came out of Africa. Some recent studies in mitochondrial DNA-distancing seem to indicate that modern "man," actually modern *woman*, first emerged about 225,000 years ago in what is now the Sahara desert of Africa[1], but earlier studies by scientists who actually invented the DNA-distancing method suggest that the primary races of humanity had dispersed and differentiated as long ago as 300,000 years, and were perhaps descendants of slightly differing kinds of "ape-men."[2]

The important point is that at this time, no one knows what happened to propagate "modern man" all over the world. Humanity may have radiated from one geographical center, probably Africa, or not-quite-humans may have migrated to various regions and *then* became "modern man" in separate transformations which left "genetic-isolation/ geographical" or *racial* marks upon them.[3]

However, one thing is certain. Climatic disaster befell humanity in Europe and western Asia, and the world has been reaping the consequences since.

According to the conclusions of most physical anthropologists, there were roughly "modern" Europeans in Fontechevade (France) and in Krapina (Yugoslavia) *before* the onset of the last European Ice Age.[4] These people lived in Europe from about 225,000 years ago until about 100,000 years ago. No one can say whether they were "Caucasians" or not.[5]

About 100,000 years ago, the climate of the Northern Hemisphere changed; Europe and western Asia became much colder than presently. Central Asia remained about the same with a seasonally extreme "continental climate," while the northern areas of the Far East of Asia (and Alaska) actually became warmer.[6]

Modern humanity of Europe (the Fontechevade and Krapina people) fled southwards, probably into Saharan Africa, but some Europeans (the "Neanderthals") stayed behind to adapt culturally and physically to the cold.

The origins of the Neanderthals are, naturally, in dispute; some anthropologists lean toward the theory that they were a uniquely European development, while others think that they may have come into Europe from the East (but without pointing to any known human population from which they could have come). It seems fairly certain that, before the last European Ice Age began about 100,000 years ago, the Neanderthals co-existed with more modern-looking people represented by the Fontechevade and Krapina people. *The Iceman Inheritance* discusses the various theories of Neanderthal origins in detail.[7] What is important here is Neanderthal physical appearance and Neanderthal psychology.

The basic fact about the Neanderthals is that they were a cold-adapted kind of man and their physical adaptations became more extreme as the arctic conditions of the Wurm I glacial period intensified. In general, they adapted to the cold in exactly the same way that other arctic mammals have, and an important part of this is what zoologists call "Allen's Rule": arctic animals tend to adopt a spherical shape; they are plump. This is because a sphere contains the maximum possible internal volume with the least possible surface area of any geometric figure.[8] Therefore, arctic animals tend to be plump so that the body heat is conserved; it is not radiated away by excessive surface area. The Neanderthals followed this common rule of arctic adaptations. They became progressively more stocky and squat, creatures of short stature but very massively built.

Neanderthals may have been rather primitive looking, but they were not stupid. Their average cranial capacity exceeded the average for modern man[9]; they had large brains and were very intelligent. This big brain gave them a big head, and a big head on their short and stocky bodies gave them a physical appearance that would be called "grotesque" by today's standards. Their faces were about twice as large as modern faces, but on a creature only about 5'7" tall at most (for a large male); some female Neanderthal skeletons indicate women who were, in life, only about 4'11" in height.[10] The shape of Neanderthal heads was distinctive; they had a long skull which over-hung the back of the neck considerably, and a low, broad skull which tended to over-hang their ears; the skull usually showed two definite "dips" down the center-line at both major cranial sutures (the meeting of major skull bones).[11]

But the most distinctive feature of Neanderthal heads was their noses. Neanderthal noses were very large with equally large nostrils and protruded from the face like a kind of blunt beak. The rest of the face protruded forward to "catch up" with this formidable nasal development. This large nose was a special adaptation to the cold. It took air in as far as possible away from the brain and a complex system of Neanderthal sinuses further warmed this air as it approached the brain.[12] Neanderthal feet and hands were massive and powerful, but fully human. A Neanderthal footprint, preserved in the clay of an Italian cave, reveals a very short, broad foot with a plantar index of 1.61; translated into modern shoe sizes, this is an EEE and is not normally manufactured.[13]

The Iceman Inheritance argues that physical adaptations demanded by Ice Age climate conflicted with sexual adaptions that had already been evolved in humanity in order to absorb the aggression caused by the "time as territory" conceptualization. With the Neanderthals, some of these sexual adaptations were almost wholly erased, others were partly nullified and still others were grotesquely over-exaggerated in an attempt to retain their function on an extremely cold-adapted body type. This must be explained in sufficient detail, because of its crucial relevance to Caucasian aggression in general and to Jewish aggression in particular. A fuller exposition is contained in *The Iceman Inheritance*.[14]

Arctic animals are not only adapted by becoming plump in accord with Allen's rule, they are also adapted by having warm fur. *The Iceman Inheritance* proposes that Neanderthals had to have had a thick growth of hair, perhaps even approaching furriness.[15] This, of course, compromised one of the basic human sexual adaptations: loss of hair in the interests of greater tactile stimulation. Humanity, as a whole, can be considered a sort of "naked ape," as Desmond Morris would have it, but Neanderthals must have been less naked than other major groups of people. Nonetheless, I think that the evidence indicates that "Nature" did what she could to preserve as much tactile stimulation as possible among the Neanderthals. While all Neanderthals must have been hairier than other major groups of humanity, the males must have been much more hairy than the women. Neanderthal men, who had to leave their caves in order to hunt for food (and the evidence shows that they were very adept hunters), had to be reasonably mobile. Therefore, they must have been a little less plump than their women, but must have had a great deal more fur.

On the other hand, Neanderthal women did not have to leave their caves so often, and indeed must have been largely confined to them by domestic and child-rearing duties. In this case, some tactile stimulation could be preserved by insulating the women with fat rather than with fur, making them less hairy than their men, and preserving skin that was a bit more "naked" in the interest of tactile stimulation. Neanderthal women's buttocks and breasts probably became *very enlarged* to preserve their stimulatory functions on an already plump body type. If things went along these lines, and many existing Ice Age "Venus figurines" show precisely this female body type, then certain crucial sexual adaptations were "preserved." However, they were preserved *at the cost of making Neanderthal sexes very different in appearance* and at the cost of emphasizing normal sexual differences in life-style and social role. Since sex provokes a great deal of aggression in all human males of whatever "major genetic group," the unfortunately extreme differences in sexual appearance and role among Neanderthals must have induced an even more extreme "desire-repugnance," or "intercourse-violence," cycle among Neanderthal men. This must have resulted in alternating moods from "sexual desire and extreme sentimentality" to "rejection, aggression and violence."

It is known that Neanderthals did have sensitivity to beauty. At the Neanderthal burials at Shanidar, pollen analysis of debris proved that flowers chosen for decorative appeal were placed beside the deceased.[16] This seems to indicate a capacity for gentleness. Yet, among the same people, Trinkhaus and Stewart proved an inordinate number of injuries caused by violence. To me, this indicated a Neanderthal psychology ruled by alternating bouts of aggression and extreme (maudlin?) sentimentality. This highly maladapted psychosexual orientation would, of course, be exhibited primarily by male Neanderthals since the basic ambivalence of "territorial assertion in time" manifests itself through male animals.

Because of their covering of hair or fur, Neanderthal men lost much tactile sensitivity and they could not have valued it very much. At the same time, in accordance with Allen's Rule for arctic adaptation, Neanderthal penises must have been smaller than those among other major human genetic groups. This resulted in less tissue area for nerve synapses, less sensitivity. At the same time, though, Neanderthal women had very large hips because they had to give birth to babies with very large heads. Neanderthal female pelvic bones show that they were wider in diameter than the norm for modern Europeans[17]—and modern European pelves can be an inch wider in internal diameter than the norm for other major genetic groups.[18] This means that Neanderthal vaginas were probably quite large and this, combined with a small Neanderthal penis, could not have contributed to a happy situation in the Neanderthal sex life.

Neanderthals had Europe to themselves during the worst of the Wurm Ice Age, from about 100,000 years ago until about 40,000 years ago. By then, they had developed very distinctive physical traits [19] (and, one suspects, equally distinctive psychosexual traits) and they must have been fairly numerous, in spite of the harshness of their climate, because of known geriatric ratios among skeletal material.[20]

About 40,000 years ago, a temporary, but intense, warming of the European climate known as the Gottweig Interstadial occurred.[21] People who had formerly lived in Europe and who had been waiting out the Ice Age in Saharan Africa, descendants of the Fontechevade and Krapina people, migrated back into Europe. One supposes that these returning migrants fought with the resident Neanderthals in some places, but also inter-bred with them. After about 8,000 years, or about 32,000 years ago, the climate turned cold again and the "modern European" people retreated once again. Europe was left to the Neanderthals and to whatever mixed populations of Neanderthals and immigrants who preserved sufficient cold adaptations to survive Wurm II. These were "Neanderthal-like" people, by this time, and Carleton Coon gave them the label "Neanderthal-Caucasoids."

Now, some climatologists think that this second cold snap, which happened about 32,000 years ago, lasted until about 20,000 years ago—or for 12,000 years. Then, about 18,000 B.C., there was a very brief warm period that lasted only 2-3,000 years when southerly people migrated into Europe and once more inter-bred with the Neanderthals and mixed Neanderthal-like people. Then, about 15,500 B.C., the cold deepened again (Wurm III) and Europeans went back into caves, again preserving and exhibiting Neanderthal-like physical traits although now much-diluted ones. Finally, about 8-10,000 B.C., the Ice Age was over and the glaciers began to make their final retreat. Southerly people began to migrate into Europe once more and inter-bred with the remaining Neanderthals as well as with the Neanderthal-Caucasoids who already represented a mixed population.

However, not everyone agrees with this scenario. One of the greatest climatologists, Cesaro Emiliani, believed that Wurm I lasted from about 80,000 years ago until about 50,000 years ago. Then there was a warm period, permitting migration of southerly people into Europe from about 50,000 years ago until about 40,000 years ago—a matter of 10,000 years. After that, the climate got progressively colder to culminate about 20,000 years ago in glacial conditions that were as cold as ever before (Wurm II), and then 15,000 years ago the weather began to warm and has been warm ever since. For Emiliani, there was no warm snap 18,000 years ago and no Wurm III.[22]

But it hardly matters. The point is that Neanderthals ruled Europe exclusively during the Ice Age; either once or twice there were warm periods when returning peoples arrived in Europe and inter-bred with the Neanderthals; once or twice when the weather turned cold again, Neanderthal cold-survival traits became selected again (and then the end of the Ice Age about 10-15,000 years ago or less when southerly peoples began to migrate into Europe and mix with Neanderthal and Neanderthal-Caucasoid populations). Modern Europeans are the result—modern "Caucasians," white men.

Now, it used to be thought that modern man, labelled a "Cro-Magnon Man," came into Europe and the more primitive Neanderthals "died out." This has been repeated, ad nauseum, in books by popular writers of history and anthropology who do not have the foggiest notion of what they are talking (or writing) about. The skeletons found at Cro-Magnon in France represent *one* of *many groups* of people that migrated into Europe whenever the weather allowed them. When Neanderthals and newcomers met, they did what human groups have always done: they fought and mated. Sometimes, more or less distinct clans of Neanderthals and Cro-Magnons co-existed within miles of each other, and for hundreds of years. They gradually melded, to a greater or lesser extent, depending on the time and geography, but the Cro-Magnons did not "supplant" the Neanderthals. Neanderthal genes are alive and well, and dwelling within every European. This does not need to be emphasized, at least not for physical anthropologists who have known these facts for many decades. Yet, it may come as a shock to readers (and writers) of popular history and anthropology books that are based on naive anthropological myths of the era 1910-1920 when the Cro-Magnon adoption was in full swing.[23]

On June 26, 1991, Dr. Christopher Stringer of London's Natural History Museum gave a press conference about the just-completed excavation of a French site where Neanderthals and Cro-Magnon peoples co-existed, *within miles of each other*, as late as 10,000 years ago—about 8,000 B.C.—or about the time when the foundations of Jericho and Catal Huyuk were being laid (the first known towns). This should not come as a shock to anyone. Why should fairly pure communities of Neanderthals in France not exist 10,000 years ago? Nonetheless, the more sensational tabloids gave prominence to this mundane discovery with loud expressions of amazement that "brutish" Neanderthals, who may have been *cannibals*, lived so late in human history.[24]

As the Ice Age glaciers retreated, and as the weather warmed up, many groups of people migrated into Europe and western Asia. Some of these people were former Europeans (Fontechvade and Krapina cultures—who may have been "white" Caucasians), but others may have been Blacks and Capoids; but most were probably well-mixed hordes of all three. They inter-bred with the resident Neanderthals wherever they all encountered each other, and the very variable population of Eurasia ("Caucasians") is the result.

However, when possible it seems that many Neanderthal clans migrated up into nearby mountains where the altitude preserved a cool climate in summer and a truly glacial one in all other seasons (just the weather they were accustomed to). In mountainous parts of Europe and western Asia, the modern population is short and sturdy (eg., Bavarians, Georgians of the Caucasus, some Swiss) and these people are more "Neanderthal" than "Cro-

Magnon." The Basques, in their Pyrenees, are another example. Their very distinctive language may, in fact, be a dialect of pure Neanderthal speech uncontaminated by southerly linguistic infusions, since the Basque language apparently has no affinity whatsoever with any other known language or even with any major language group.[25]

In Europe itself, no matter how far up the mountains Neanderthals may have trekked, there was a good deal of inter-mixing with incoming migrants and a great deal of inter-marriage up to the present so that Neanderthal traits in some mountain populations, while obvious enough, are much diluted. Things were much different in western Asia—or, that undefined geographical region where "Europe" and "Asia" meet. Neanderthals also lived there and there are many imposing and interrelated mountain chains, but this region was a kind of backwater in terms of migration routes. Newcomers from Africa arrived later than in Europe itself, and in far fewer numbers. Therefore, this is where one would expect Neanderthal traits to survive in purer form. And this is precisely the case. Peoples of the Caucasus mountains and the surrounding Russian steppes very often exhibit such pronounced Neanderthal traits that the temptation is to suppose that they *are* Neanderthals whose physical characteristics have not been diluted by inter-marriage with migrants, but simply by the fact that a warmer post-Ice Age climate has allowed some gradual flexibility in the expression of genes departing a bit from the rigorous Neanderthal Ice Age blueprint, genes that would have been suppressed, "selected out," under colder conditions.

Carleton Coon devoted much effort in his *The Origin of Races* to the task of correcting the popular misconceptions about Neanderthals. The extreme, truly Ice Age

Neanderthals were grotesque by modern standards but, even so, there is not one Neanderthal bone that cannot be matched among living Europeans. Of the sturdy and broad Neanderthal feet, Coon said: "They looked like Russian feet, rather than English feet."[26] Of the equally massive hands and arms, Coon said that "they were not notably different from the hands and arms of hard-working Europeans today."[27] Among modern Caucasus people, some Slavs—and many Jews—one can see pure Neanderthal noses just like the plaster reconstructions of Neanderthal faces (based on actual Neanderthal skulls) reposing in the University Museum at the University of Pennsylvania where Dr. Carleton Coon was Professor Emeritus of Anthropology.[28]

Neanderthal brains were larger than the modern norm. Neanderthal caves have relinquished bones inscribed with lunar notations, showing that Neanderthals could count days of the month more than 50,000 years ago. Remarkable clay sculptures of a bison and several bears have been discovered in Neanderthal caves and Neanderthals apparently played a game, like darts, in which clay balls were thrown at a *graduated* round target on the cave wall.[29] This means that the graduations must have had different numerical values and that scores were kept. Indeed, from the scant data that survives, it seems that Neanderthal brains were not only very well developed, but had a decided "spatial-mathematical" bent. One might suspect that people with strongly Neanderthal characteristics would excel at music, mathematics and disciplines requiring spatial (three-dimensional) conceptualization: sculpture, engineering. They do not seem to have had much sensitivity to color; no Neanderthal paintings have been discovered, perhaps they created none. Painting seems to have been the specialty of those people who happened to migrate into Western Europe from Africa. These people may have been returning "Europeans" of the Fontechevade and Krapina cultures who had acquired a strong Capoid admixture while resident in Saharan Africa, or they may have been Capoids (the celebrated French and Spanish cave paintings resemble Bushman cave paintings in terms of style and content).

It is one of the ironies of human pretension that, just shortly after Europeans had accepted the indignity of having "evolved from the apes" (as Darwin was misrepresented) and had repugnantly accepted the fact that "brutal" Neanderthal Man had occupied Europe in the dim past, Cro-Magnon skeletons and cave paintings of Altamira (and elsewhere) were discovered. Cro-Magnon was enthusiastically welcomed as "modern man" and our true ancestor because he apparently had a chin, and Neanderthals did not.[30] The paintings were taken as proof that Cro-Magnon's lineage, leading to modern white man, had evolved sensitivities.

And this occurred, at least partly, because the experts—who discovered Cro-Magnon and the cave art—were from France, a country that justly took pride in its tradition of famous *painters*. So, Cro-Magnon and the cave art were eulogized; Neanderthal was despised as brutish and killed off so that it could be argued that Europeans descended from Cro-Magnon and the marvellous cave painter . . .but . . .no one enquired whether the cave painters could *count*. Now, they probably could count (although we have no proof), but the fact is that the poor, despised Neanderthal was counting, playing "darts" and molding three-dimensional statues 30,000 years earlier. No one appreciated *that*, few appreciate it now. Neanderthal seems to have had a "mathematical-spatial" type of intellect, rather than a "visual-sensory" one, but Neanderthal's abilities have proven no less valuable to human material development than the gift of the cave painters.

Since the Eurasian mountains were far removed from the major migration routes at the close of the Ice Age, and since strongly "Neanderthaloid" populations exist today in the Caucasus region, it seems reasonable to wonder whether really pure, genuine Ice Age Neanderthals still survive. There can be little doubt that the mountain trolls and forest gnomes of European folklore are not-so-ancient memories of fairly pure Neanderthal remnant populations, but could these trolls and gnomes yet live in remote Eurasian mountain ranges?

Between 1900 and 1910, the Russian (then "Imperial") Academy of Science sent a young zoologist, Vladimir Khakhlov, to do a survey of animal life in remote areas of "Eurasia" (where Europe and Asia meet). He was sent to the inter-locking chain of mountain that formed the southern border of Imperial Russia and which, until just recently, also formed the southern boundary of the Soviet Union. The mountains comprise the Caucasus group between the Black Sea and the Caspian Sea, then the Elburz chain that wraps around the southern end of the Caspian Sea and proceed eastward across northern Iran and into Afghanistan, the Hindu Kush range that occupies most of Afghanistan, the high Pamirs in the jumbled border region of Pakistan and China, the Tien Shan mountains in what is now mostly Sinkiang Province of China, then across the scarcely 100-mile wide strip of steppe called "Dzungarian Gate" linking the west with the Far East, to the mighty Altai Mountains

Figure 1.

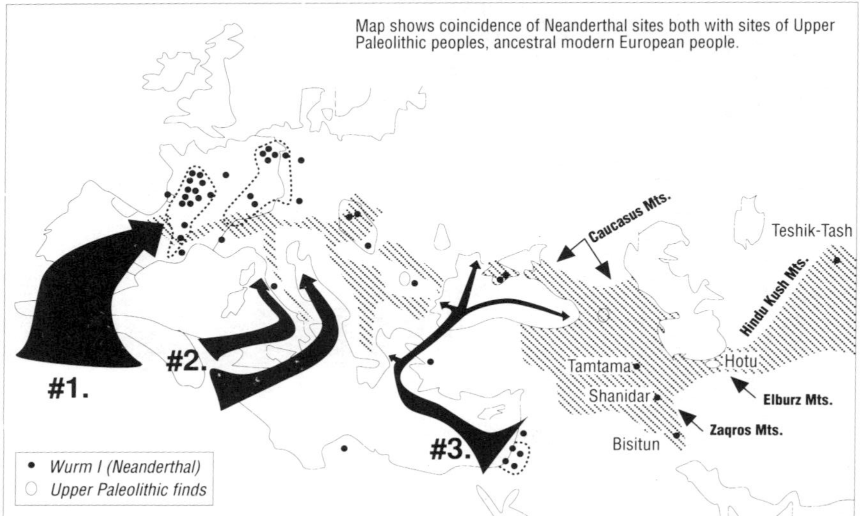

Post-Glacial Migration Routes Into Europe

"Upper Paleolithic: is the anthropological term given to the mixed peoples who inhabited Europe after the initial phase of the last Ice Age (Wurm I). They were a mixture of incoming migrants from Afroasia and resident Neanderthals. In this map, based partly on Carleton Coon's diagram (published in his *The Origin of Races*), it is clear that the concentration of "Upper Paleolithic" Europeans included former Neanderthal sites in Western Europe. I have added to Coon's work shaded areas representing mountains. Evidence from excavations to date indicates three major migration routes of post-glacial times. Most important was the route from North Africa into Iberia across the Straits of Gibraltar. Second in importance was the route from modern Tunisia that island-hopped across to Italy with two coastal migrations straddling the Italian mountain spine. The third route, and apparently the one involving the fewest incoming migrants, followed the eastern Mediterranean coast, through the Aegean and Greek Islands, and then on into the Black Sea coastal regions. It is clear from the distribution of "Upper Paleolithic" sites, that these incoming Afroasian migrants generally avoided the higher mountains of Europe initially, mountains that were doubtless refuges for surviving Neanderthals for a long time. In Eastern Europe, the resulting "Upper Paleolithic" population retained stronger Neanderthal traits for a much longer time than was the case in Western Europe. The relevant factor seems to have been the ruggedness of the mountains, the extent of the mountains chains in the Caucasus-Elburz area, the concentration of resident Neanderthal people — and, therefore, the much longer time required for the trickle of post-glacial migrants to influence the native Neanderthal genetic type. At least, this would seem to be a fair and objective interpretation of the available data.

which continue (under various names) northeastward to the Arctic. Khakhlov was sent to examine animal life north and west of all these mountains, which form a kind of "rim" all around Eurasia and the "Caucasoid racial cradle."[31]

His duties included trying to estimate how many European bison, or "wisent," might still live in this vast area, if some species of ancient horse (like the then-known Polish tarpan) might be found there and to estimate the population of tigers and leopards in Eurasia.[32]

Khakhlov completed his formidable task in exemplary fashion, considering that this mountain rim is no less than 3,000 miles in length, features some of the most rugged geography in the world and is inhabited by various tribes (of Caucasians) who did not like the Russians in 1910 any more than they liked more recent Soviet intruders. There were many places visited by Khakhlov where no Russian "European" had ever been before (and which, one suspects, no Soviet commissar has ever seen). The mountain ranges, enclosing the lowland steppes like a very thick rim of a monstrous bowl, were utterly unknown territory in 1910, and the situation is only a little better today. In 1923, for example, a Canadian anthropologist named Jill Crossley-Pratt claimed to have discovered a remnant of ancient Chaldeans living in an isolated mountain valley in the Pamirs.[33] In the late 1940's, Anglo-American military units came across a Greek-speaking "tribe" somewhere in the Hindu Kush of Afganistan, a tribe that still wore visibly "Classical Greek" togas (actually "chitons" in Greek); they turned out to be remnants of Alexander the Great's army![34] In 1954, Soviet aerial reconnaissance discovered a whole new *mountain range* just west of the Altai jumble and therefore in Eurasia.[35]

Khakhlov did his survey of animal species and estimated populations, but, early on, he came across literally hundreds of reports of an "animal" so strange and intriguing that this creature actually became the emphasis of his massive report. Mountain villagers of the Caucasus called it a *kaptar*; further east in the Hindu Kush and Pamirs it was called a *guli-yavan*; while the people of the Tien Shans and Altai mountains on either side of the Dzungarian Gate called it an *almas*. The names might change from place to place, but the descriptions Khakhlov gathered were remarkably consistent. It was a short, but very massively built, human being. It had a big head. Females had huge hips and huge, pendulous breasts. Females and young ones travelled in fairly large groups, but the males seemed to be solitary. They could range in color from black-furred to almost blond-furred, but the most common color was a sort of rich mahogany. Their fur, or hair, was very profuse (except on the breasts of the females) and was usually straight. However, the hair of the head was often curly, like a woolly mahogany fleece. They had large, protruding faces and very large noses with large, elongated nostrils. They lived in very remote valleys high up in the mountains, but they were not at all rare. Most of the mountain folk had seen at least one or two during their lives. Khakhlov was shown shacks in which one or two captured creatures had been imprisoned, and a chain-and-collar which had shackled an unfortunate female who had been used, for some years, as a sexual toy by the men of one Kirghiz village. Several had been killed.

These *kaptar/guli-yavan/almas* had a primitive form of language. They wore crude clothing on occasion, and would carry on trade by barter with remote mountain villages. They could inter-breed with modern humans, and the child was usually very smart. One boy, who had been born among *almas* of an *almas* woman when his father had become separated from a caravan in a high pass, later (after he and his father escaped) joined the Lamin-gegan monastery. According to Professor Rinchin of the University of Ulan Bator, this incident was still remembered as late as 1957. Rinchin says: "The pupil achieved scholarly fame and was known as the son of an Almasska."[36]

Since Khakhlov collected several hundred reports, he concluded that some sort of "wild man" lived in the mountains of Eurasia. He completed his report in 1914 and submitted it to the Imperial Academy, but this was on the eve of World War I and the Revolution that was to follow in 1917. Khakhlov's report was shelved, and forgotten.

Then, in 1951, mountaineer-photographer, Eric Shipton, of the Everest Expedition, brought back his famous photograph of some large, not-quite-human footprint in the Himalayan snow. Shipton's guide, a Sherpa, claimed to know all about the wild men of the high mountains and called it a "yeti." These stories of wild mountain men, and this word "yeti," had been coming out of the Himalayas and Pamirs for years. About a decade previously, a columnist for the *Calcutta Statesman*, Mr. Henry Newman, had stated for his readers that *yeti* meant "abominable snowman."[37] The Western press picked up on this name, associated Shipton's very real and very puzzling photographs with this "abominable snowman"—and most people in the Western World have heard about it.

The Soviets also heard about it and they were not happy. Their official position was that this "abominable snowman" was a purposeful myth invented by Western intelligence agencies in order to justify sending scientific expeditions into the Pamirs and Himalayas—expeditions whose sophisticated equipment were actually intended to spy across the Soviet border. Several stories of this ilk appeared in *Pravda* during the early 1950's.[38] Then, however, someone discovered Khakhlov's massive and now dusty report which had been unknown and ignored for almost forty years.

The Soviets changed their tune and undertook proper scientific investigation of the situation while the Western press continued to regale readers with fanciful "abominable snowman" stories, and most Western scientists contented themselves with ridiculing the whole affair. The Soviets immediately dispatched four properly equipped expeditions: one to the Caucasus, one to the Elburz-Hindu Kush, one to the Pamirs, and one to the great Altai barrier that separates the Mongoloid domain from the Caucasoid one. The entire effort was under the direction of Drs. B.F. Porshnev and A.A. Shmakov, members of the U.S.S.R. Academy of Sciences. These four expeditions went precisely to the places Khakhlov had visited almost half a century earlier. The reason for this was that in the intervening fifty years, Neanderthal bones had been found at the following places: Starosel'e in the Crimea in 1952; Kiik-Koba, also in the Crimea, in 1924; Teshik-Tash in Uzbekistan in 1938; Musa Dagh in Turkey in 1954; at Bisitun (by Carleton Coon) in Iran in 1949; and at Shanidar (by Ralph Solecki) in the Zagros-Elburz mountains in 1953.[39]

Quite obviously, Khakhlov's "wild men" could not have been inspired by these discoveries because all these Neanderthal remains had been discovered long after Khakhlov had filed his report. And, since Khakhlov had submitted hundreds of reports of Neanderthal-like creatures just a stone's throw from where actual Neanderthal remains had subsequently been found, it seemed more than possible that some relatively pure Neanderthal remnant populations might still exist in these remote mountain areas—or, at least, might have existed until very recently.

The four Soviet expeditions accumulated much the same sort of information that Khakhlov had gathered fifty years before: literally thousands of consistent descriptions and eye-witness reports, drawings, a few bones and footprint casts (Professor Porshnev found a *fresh footprint* in the mud beside a Kirghiz river, a print almost identical to the Neanderthal footprint from the Italian cave mentioned earlier). Expedition reports were published in four volumes by the Soviet Academy of Sciences between 1958-1960 under the title *The Reports of the Special Commission to Study the Snowman of the U.S.S.R.* Since the publication of the original reports (about 5000 pages in total), there have been detailed follow-up articles in *Tekhnika Molodyozhi*.

Although the Soviets did not capture a living *kaptar*, they did obtain much hard and soft data, including a dissected forearm with muscles still attached which "is almost certainly Neanderthal" according to Professor Teizo Ogawa (1957)[40]. At least two Soviet expedition personnel saw kaptars through binoculars and a Soviet Army medical doctor actually examined a living one in Dagestan (1941) and had submitted a detailed anatomical report which was unearthed by the Soviet commission.[41]

Their existence, at least up to the very recent past, cannot reasonably be doubted. And, from what is known of human evolution during and since the last Ice Age, these *kaptars* (and kindred creatures) can only be nearly pure

Neanderthals. These creatures must have migrated higher up into the mountains as the climate warmed, seeking the kind of weather for which they were adapted. Since these mountains happened to be far removed from the post-glacial migrations routes of other peoples toward Europe and western Asia, some Neanderthal populations did not intermix very much, *or at all*, and so preserved their genetic purity for a long time, perhaps too long. When modern man finally inhabited the foothills of these mountains, there was too much cultural and physical difference between the two groups to permit assimilation. These "mountain Neanderthals" had remained too distinctive because they sought, and continued to live in, climatic conditions which had made them distinctive in the first place. In the foothills there was more assimilation over thousands of years. Some of these mixed people would have fairly diluted Neanderthal characteristics, but other groups might have preserved fairly strong Neanderthal characteristics, depending on migration routes, geography and circumstance. But the point of all the foregoing is not primarily to prove the continued existence of the nearly pure Neanderthals in the Eurasian mountains. The point, really, is to emphasize how remarkably the descriptions of *kaptars* concur with: (1) reconstructions of Neanderthals, based on skeletal remains, undertaken by recognized experts in physical anthropology (such as C.S. Coon's reconstructions at the University of Pennsylvania); (2) depictions of Ice Age women (at least) in existing Ice Age art (the so-called "Venus figurines"); and (3) the physical "so-called pen portrait" of some Neanderthal physical traits, not indicated by reconstructions or Ice Age art.

If these *kaptar* descriptions and traditions are to be

trusted at all, they not only lend support to the above, but add to our picture of Neanderthals; their fur color was predominantly and commonly a deep red. Their head hair (or fur) could be curly ("like a woolly fleece"). They could speak a "primitive language." Some Soviet scientists have maintained that Neanderthals might not have been able to speak at all because the structure of their neck vertebrae impinged upon potential larynx development.[42] But the *kaptar* traditions insist that they can speak, but perhaps only in guttural-sounding grunts and a few single-syllable sounds.

It goes without saying that these *kaptar* are also noted for their stench, but this would hold for any wild animal living under such conditions—and for any modern person living under the same conditions for any length of time. It is interesting, though, that this characteristic smell of "wild men" has been compared, more than once, to the smell of a field of potatoes.[43]

As we shall see, some of these apparently trivial details about *kaptars*, which in no way contradict our picture of Neanderthals but only add to it, become extremely suggestive when we come to discuss "Esau" of the Bible. I have placed Esau in quotes in the last sentence because most people naturally think that Esau was an isolated "throwback" born to Isaac; the *Old Testament* purposefully gives that impression. But the truth of the matter is that the Israelites were, quite literally, plagued with offspring like Esau. They are referred to as the "Hairy Ones" in the *Bible* and, according to rabbinical tradition, they were numerous enough to contest Jacob's leadership of the Hebrew clans. The Israelites made every effort to exterminate these numerous "Hairy Ones" among them, Esaus who were very frequently born to Hebrew mothers. They were exterminated, and they were called "scapegoats . . ."

III
The Iceman Incarnate (End of Eden)

At the end of the last Ice Age, major events transpired which have had a direct bearing on true Hebrew history (as distinct from Jewish myth), Western history in general, and even Biblical interpretation. So far, the term *Ice Age(s)* has been used rather blithely, without a definition or explanation of what is meant. The following digression will add greater insight to our discussion of Hebrew origins and will also suggest great geological changes in Africa in the surprisingly recent past which may have molded the development of ancient Egyptian culture and hint at the real location of Eden.

Like so many things in Western science, the "Ice Ages" were proposed to explain one fairly simple observation, and the explanation turned out to be more complex and puzzling than the original "problem." This problem was the so-called "erratic boulders" of Europe (and elsewhere). Between 1800 and about 1840, European geologists had noted the existence of many erratic boulders; there were huge rocks, or boulders, that were presently many miles from the rock formation to which they belonged. Some of these boulders weighed 10,000 tons or more, some were perched high up on mountain sides, and it was difficult to imagine anything that could have moved them. Georges Cuvier (1769-1832), a brilliant French zoologist and geographer, had proposed a theory of vast and cyclical earth catastrophes that occurred in the past and might happen again in the future. His idea of "the Revolutions of the Globe" supposed that because of huge geological upheavals, including volcanoes and earthquakes, the oceans would sometimes sweep over the continents in huge "waves of translation" (he called them) that would carry even 10,000-ton boulders from their origin and leave them beached high and dry many miles, even hundreds of miles, away.[1]

Much geological and fossil evidence argues for great and relatively abrupt changes, but the problem with Cuvier's idea of periodic catastrophes was that he was French and had been a respected scientist under Napoleon's regime. The British, however, had just won the Battle of Waterloo. Although this might seem to be an absurd and non-sequitur statement, it is not. The British had recently obtained military dominance in Europe after defeating Napoleon. They did not like "revolutions" and "catastrophes" of *any* kind (particularly not after Napoleon so dramatically changed European society), and their scientists sought theories that were more in accord with the English temperament. If there had to be changes at all, the English soul cried out for a slow, dignified process of change — no sudden revolutions. Sir Charles Lyell (1797-1875) did much to promote and popularize the theory called *Uniformitarianism* which opposed Cuvier's *Catastrophism.*[2] The dogma of "Uniformitarianism" argued that no geological forces operated in the past that we do not see in action today. Therefore, if we do not witness catastrophes of global scope today, such catastrophes could not have happened in the past.

Now, it has to be said right away that the doctrine of Uniformitarianism was never, strictly speaking, a proper scientific *theory* because much evidence from the past geological strata argued against it vehemently. It was more in the nature of an "article of faith" or "a manifesto of British thinking" rather than anything else because the British were simply tired of, and horrified by, the social upheaval wrought by Napoleon and his wars. They were determined not to have anything to do with upheavals *of any sort*—in society, science, the past or the future. When Wellington was finally able to defeat Napoleon at Waterloo, and British military force became supreme in Europe, then British scientific notions became dominant. The entire emphasis of 19th century science, and even science in most of our 20th century, would have been much different (less dogmatically stuffy, more adventurous) if only an aging and often-defeated Prussian general by the name of Gebhard Leberecht von Blucher had not committed an almost super-human (and totally unexpected) act of will. Badly beaten by Napoleon on the morning of Waterloo, with his forces streaming away in disorder and near-panic, von Blucher somehow decided *that he would not be defeated again!* He rallied his men, organized them, and led them back to Waterloo just in time to aid Wellington (who was also being mauled by the Corsicans). Blucher made Britain's victory possible and, unknowingly, ushered in an era of science which, while it has numerous triumphs to its credit, has also steadfastly refused to examine many facts.

The erratic boulders existed after Waterloo. But now they could be explained by something less drastic than Cuvier's "waves of translation" inundating entire continents. Louis Agassiz, a Swiss geologist, is usually credited with propounding the theory of the "Ice Ages," although he actually "borrowed" the theory from an obscure Austrian school teacher.[3] Agassiz (1807-1873) noted that the alpine glaciers in his native Switzerland had once extended much further down the mountain valleys than was the case in the 1830's when he did his initial work. At some point, the glaciers had been much, much larger and had extended for many miles further. At last, they had melted, but they had left behind heaps of rocks, boulders, pebbles and other debris which had been pushed along at the edge of the advancing ice sheets. These heaps of debris are called *moraines* and there is no doubt that they were pushed into their present position by now-melted glaciers of ice. Agassiz suggested that even erratic boulders had been rolled and pushed into their present positions by sheets of ice—although, since some of the boulders are situated many thousands of feet up on unglaciated mountainsides, the "ice" must have been very thick at one time. Many thousands of feet thick.

Agassiz went to England and demonstrated the existence of *moraines* there. It became accepted that, at one time, thick sheets of ice must also have covered the British Isles. Very quickly, geologists in other places discovered moraines in different parts of Europe, western Asia and

North America.

It quickly became clear that a lot of ice had once covered a large part of the Northern Hemisphere, in places where none exists today, and the idea of a past "Ice Age" came into being. This idea was not, and is not, exactly in accord with the doctrine of Uniformitarianism; we do not see a mile-thick sheet of ice over Britain and Chicago today, although something of the kind must have happened in the past. It was just one more bit of evidence that the theory of Uniformitarianism did not fit very many facts. The theory of the "Ice Age" was nonetheless accepted because ice presumably had one virtue; *it moved slowly!* The British ideal of *slow*, stately (even dignified) geological changes could be preserved in ice, so to speak—much preferable to Cuvier's sudden "waves of translation" which could traverse entire continents in hours. Since ice accumulates slowly, and moves slowly (as measured by existing mountain glaciers), then the "Ice Age" must have taken a long time to build up, must have endured for a long time and must have taken an equally long time to melt. It was a tempo that British souls could accept! The "Ice Age" was confidently placed *a long time* in the past.

By the era 1875-1900, geologists had discovered that there had not been just *one* "Ice Age" within the last million years or so, there had been *four*.[4] In Europe, the sequence was first, the Gunz Ice Age, followed by a warm period; then the Mindel Age, followed by a warm period; then the Riss Ice Age, followed by a warm period; and finally the Wurm Ice Age (which had at least two phases), followed by a final melting and today's climate. In North America also, four separate ice advances were discovered and these were named, not after geologists as in Europe, but after the states to which each ice sheet advanced. First, there was the Nebraskan glaciation, followed by a warm period (Aftonian Interglacial); then the Kansan ice sheet, followed by a warm period (Yarmouth Interglacial); then the Illinoian ice age, followed by a warm period (Sangamon Interglacial); and then, finally, the Wisconsin ice sheet, followed by a general warming to result in today's climate.

There were two cold periods to the European Wurm Ice Age, called Wurm I and Wurm II. Some scientists argue a Wurm III. In North America, some geologists equate the "early Wurm" in Europe with an Ice Age called Iowan, and equate only the later phases of European Wurm to the Wisconsin ice sheet.

In short, by about 1900, it was recognized and accepted that there had been four "Ice Ages" in Europe and North America within the past million years. Given the lengthy intervening warm periods, it became obvious that these Ice Ages could begin, culminate and melt in a surprisingly short period of time (as geologists count it), say, 100,000 years or so—a bare thousand centuries. Not very long for such a fundamental climatic change, and the doctrine of Uniformitarianism was rocked a bit. Having isolated four separate ice advances, scientists felt obligated to try to explain this phenomenon.

There were, and are, three major ideas about what causes Ice Ages *that are based on the assumption of a genuine worldwide cooling of the earth's climate.* These "theories" and their underlying assumption are still prevalent. First, there was the thought that the sun's output of heat energy might vary in some cycle that we have not been around long enough to observe scientifically and which cannot be explained by current solar theory.[5] Then came the idea that the entire solar system might, at intervals, pass through clouds of cosmic dust and gasses that inhibited the sun's radiation from reaching the earth. Whereas this is very possible, the clouds of dust and gas have never been identified by our optical or radio telescopes.[6] Finally, there was the idea that if the several known irregularities in the earth's orbit around the sun all culminated together at the time of northern winter, they could bring on an ice age; however, this was shown (mathematically) to be impossible.[7] Despite the fact that all three of these theories have been shown to be grossly inadequate in explaining successive *recent* ice ages, they are still offered as possibilities in our universities.

Strangely enough, few scientists have questioned the underlying assumption of all three of these theories. *Was there a cooling of the entire global climate* during the four recent Ice Ages? No one, apparently, examined this fundamental assumption in any systematic manner, but, gradually, some data emerged that tended to argue otherwise.

Soviet researchers doing Carbon 14 testing on mammoth bones in Siberia discovered that the Siberian climate was warmer during the last Ice Age than today.[8] Contents of mammoths' stomachs showed that certain tundra flowers, and dwarf willow trees, grew 500 miles to the north of where they grow today.[9] So, about 100,000 to 15,000 years ago, when Europe was in a deep freeze and Neanderthals managed to survive in Eurasian caves, the climate of Siberia was considerably warmer. Alaska was also warmer than today.[10]

Then, Professor Ting Ying H. Ma of the National University in Taipei, Taiwan, published (between 1943 and 1952) a six-volume work that must rank as one of the foremost "textbook" examples of how to conduct careful scientific research.[11] Ma studied corals from all over the world and from all geological eras. With infinite patience and meticulous accuracy, he plotted the distribution of some 300,000 separate pieces of coral on a world map—actually on several world maps corresponding to various major geological eras. Now, corals are organisms that exist only in tropical and sub-tropical seas; they cannot withstand much cold or variation in temperature. Ma discovered that corals have not only always existed (which everyone knew), but that corals have always occupied about the same-sized zone around the earth, although these zones have been in different locations around the earth at various times in long-past geological history. In short, Ma proved, without a shadow of any doubt, that the tropic and sub-tropic zones of the earth have always been just about the same size as they are today. That being so, there could not have been any worldwide "Ice Age(s)," *no cooling of the entire earth's climate.*

Ma's research on corals was confirmed by other researchers working with other kinds of tropical animals (notably, snails). If there had been a true worldwide cooling of the climate, then many kinds of tropical creatures, those very sensitive to any sort of temperature change of more than a few degrees, would have died out completely. Yet, because they still flourish in today's tropical zones, the inevitable conclusion must be that there was no worldwide cooling of the earth's climate during the so-called Ice Ages.

Some scientists, absorbing all this evidence, tried to figure out a scenario that fits the facts. One obvious idea was that the earth's axis must have shifted. Now, the axis of the earth is presently inclined at $23^{1/2}$ degrees to the plane of the ecliptic—which is to say, more or less, *to the sun's equator.*

Thus, as the earth goes around the sun, the northern hemisphere is tilted away from the sun during the northern winter and, at the same time the southern hemisphere is tilted toward the sun so that Australia has its summer between January and March. On the other side of earth's orbit, the opposite happens: the northern hemisphere is tilted toward the sun for our northern summer while Australia has winter between June and September! If, somehow, this axis shifted—tilted more, or less, or in a different way, then the polar zones would be in a different place on the earth's surface—nearer to Europe, for example, and Europe could be considered to have an "Ice Age" (it would simply be in the Arctic zone). Another axis shift, and Europe might get appreciably warmer while someplace else would get appreciably colder and have an "Ice Age." It was thought that a few of these "axis shifts" could explain the alternating cold and warm periods that have dominated the last million years of Eurasian and North American history.

This seemed like a good idea—although the doctrine of uniformitarianism no longer applied—but the snag was, first, that physicists easily proved that no amount of energy from within the earth itself, no imaginable earth cataclysm, could tilt the axis. Too much energy was required. The axis could only be tilted by some celestial object of great size coming near to the earth.[12] Indeed, it is thought that the earth's present axis tilt of 23¼ degrees may have been caused when the solar system was formed over four billion years ago. As the earth was thrown off as a ball of sun-stuff, either the sun itself or one of the larger planets then a-forming, passed close to the earth and imparted the axis tilt.[13]

This axis tilt is responsible for seasons. If this tilt had ever changed significantly in the recent or more distant past, then the fossil record would show some havoc with the seasons. But paleontologists were able to show that as far back as the Cretaceous (which ended 72 million years ago, according to the latest estimates)[14] some trees seasonally lost their leaves, just as they do today.[15] So, there have been seasons as far back as 72 million years. The axis had not tilted appreciably more or less in all that time (although it may have shifted somewhat), and axis tilt could not be held responsible for the recent European/North American Ice Ages.

Then, around 1950, a mechanical engineer named James H. Campbell proposed a very elegant theory to explain "relatively frequent" Ice Ages (in geological time-scales). This theory fits all of the facts (and explains a lot about geography) and was the only theory by someone else ever endorsed by Albert Einstein, but it is largely unknown among contemporary professors and it is rejected and ridiculed by those few who have heard of it.[16] Campbell's theory can be summed up as follows . . .

First, it is known that the earth has an equatorial bulge of about 13 miles, which is to say that the earth's diameter at the equator is 13 miles longer than the earth's diameter from pole to pole. This is because the earth's substance is "flung out" by centrifugal force at the equator as the earth rotates at about 1000 mph around its axis. This "bulge" around the equator actually stabilizes the position of the earth's surface crust.[17] This thin crust is composed of large "plates" (tectonic plates, significantly mapped and explained by the Canadian geophysicist, J. Tuzo Wilson)[18] that just barely cover a "mantle" of molten *rock* and lava (the true "core" of molten *metals* begins many miles closer to the center of the earth). The crust can slip rather easily over this molten mantle *without affecting the tilt of the axis at all.* What would happen is that another expanse of the earth's surface would be carried into the polar zones while, simultaneously and equally, the geography that had been within the polar zones would be carried into warmer regions. Only the equatorial bulge holds the crust in place.

What could cause the crust to move? Campbell said that the polar ice caps themselves could do this, and do it on a more or less "regular" basis, but with a frequency and amount of crustal shift that depended on some complicated factors and would be difficult to predict. Let's follow a "Campbell cycle" and see what happens. . .

Surface geography is shifted into the two polar zones (north and south). The old ice caps have, by definition, been shifted into a warmer zone somewhat nearer the equator and so they begin to melt. New ice caps begin to form, at the same time, on the earth's surface which has just been shifted into the polar zones. According to Campbell, it would form in a certain way; at first, snow would begin to fall at the pole; a lot would fall and it would begin to compress itself into actual ice. Then, at some point, the mass of ice would begin to flow outward all around the edge of this central ice core, but it might be hampered (or assisted) by the terrain under the new ice cap.[19]

It matters a great deal whether the new areas of earth in the polar zones are land or ocean. Ice would form much more slowly, and never be as thick, if a pole (or both poles) were located on an ocean. This is because oceanic currents from the tropics are continually reaching the polar zones; they would warm the water in the polar zone enough to inhibit the growth of a massive ice cap.[20] But if a pole (or both poles) happened to be on land, the ice would form much more quickly and get much thicker because moderating oceanic currents could not act directly on the ice. The world today demonstrates this phenomenon. The Arctic is located in an ocean, and so the Arctic ice cap is much smaller and less massive than the Antarctic ice cap. Antarctica is actually a series of large islands beneath the ice, islands which insulate the growing ice from warming equatorial currents. As a result, the antarctic ice cap is composed of *six million cubic miles of frozen water.*[21] This is a huge mass and weight of water.

Campbell further argued that new ice caps, whether formed over water or land, would inevitably develop so that their centers of gravity would be somewhat offset from the actual pole. On an ocean, quirks of current would act to ensure this; on land, geography would act to make an off-center ice cap. Again, in the real world of today, this is the case: the giant Antarctic ice cap's center of mass is some 300 miles away from the pinpoint of the South Pole.[22] The Antarctic ice cap is off-center by 300 miles. This huge mass is badly out of balance as it revolves at high speed around the South Pole. It is like an out-of-balance tire on a huge scale.[23]

Campbell's theory proposed that, sooner or later, one or both of the earth's ice caps would become increasingly unbalanced. At some point, the off-balance energy would be enough to cause a crustal shift. The out-of-balance ice caps would tend to shift toward the equator because of centrifugal force. When the mass of these ice caps exceeded the strength of the earth's crust at the equatorial bulge, the crust would split at the bulge and allow the entire crustal surface to shift in order to relieve the strain of the off-balanced ice caps.

Now, when Campbell first proposed this theory back in 1950, it was thought that the earth's crust was "one unit." Although Einstein endorsed Campbell's theory in 1953, subsequent calculations seemed to prove that the strength of the crust was more than sufficient to withstand the off-centre force of any imaginable ice cap. For this reason, Campbell's theory was ignored even before it became well-known. Since 1953, however, the geophysical model of "plate tectonics" seems to have been proven without doubt and is universally accepted. This model of the earth involves essentially unconnected "pieces" of the earth's crust slowly floating and jostling each other as they drift over the molten mantle. The edges of these giant plates jostle against each other, and crumple to make mountains where plates collide (such as the Himalaya Mountains, where the Indian plate is crashing into the Asian one). Sometimes edges are forced down into the mantle to be re-melted into magma while in yet other places, magma oozes up from the mantle to become solidified into new crust (as in the Mid-Atlantic ridge). All this allows for a certain amount of "continental drift," but now much different in concept from the idea originally proposed by the German biologist, Alfred Wegener, who, by the way, first suggested his idea as an explanation for the "Ice Age!"

> He (Wegener) thought of the ocean bottoms as composed of soft, amorphous viscous rock. From a vast amount of fossil evidence of the plant and animal life of the past, he imagined that he could reconstruct the paths of the continents over long periods of time. He proposed to explain the mystery of the Ice Ages by this theory. He suggested that during the last Ice Age in the Northern Hemisphere, Europe and America had

> lain together near the North Pole, but since then they had drifted away from the pole and each other.[24]

It is clear now that although the continents—or tectonic plates—do "drift," they do so much too slowly to explain the last Ice Age.[25]

However, to return to Campbell's theory, it seems that tectonic "cracks" between the major crustal plates can weaken the strength of the entire crust considerably. And, if major tectonic cracks *happen to be* oriented from pole-to-pole, and if one or both of the poles *happen to have* a huge and off-center ice cap, then the centrifugal force of the ice cap *is more than sufficient*[26] to cause a shift of the crust over the equatorial bulge. One of the tectonic cracks will widen at a point where it is already weak, and allow the whole crust to slip over the mantle. The old ice caps are suddenly moved out of the polar zones and begin to melt; new ice caps start to form.

That's Campbell's theory. It makes a great deal of sense concerning geological events in the very distant past, as well as in the much more recent past. Many hundreds of millions years ago, as Ting Ying H. Ma's work with corals proved, the crust was oriented so that both poles happened to be in oceanic regions.[27] Crustal shifts were, therefore, infrequent and of minor amplitude. Being over water, the ice caps never grew very large because of continual moderation by tropical oceanic currents. For many millions of years, even tens of millions or hundreds of millions, crustal shifts were minor.[28] Eventually though, gradual migration of the crust brought about the present situation, which has characterized the past few million years in general, and the last million years in particular. The poles are arranged so that polar zones encompass random and unequal expanses of land and water; even "worse," the major tectonic fractures happen to run roughly from pole to pole. The Mid-Atlantic ridge, the largest tectonic fracture in the world, runs from pole to pole in a very literal sense. Other great tectonic fractures are now also oriented in a north-south direction (such as much of the North America/Pacific plate, the African/Indian ocean plate, etc.).

This means that ice caps of a seriously "off-balanced" nature can build up on one or both polar zones in only 50,000-70,000 years, perhaps less. Then, the tectonic plates and fractures happen to be arranged in a way to allow frequent splits and shifts in the crust. So, the theory of James H. Campbell very neatly explains everything.

Campbell had already thought of the idea that when the crust split over the equatorial bulge in one place (say, going "up" the bulge), it would have to crumple together in another place (going "down" the bulge). This alternate splitting and crumpling would be roughly in a straight line, *in the direction of the last crustal shift.*[29] Where the crust had been forced to split, Campbell realized that there would be shorter fractures running at nearly right angles to the major fracture; and, in a similar way, there would be "relieving crumples" at right angles to the major crumpling. Any glance at a map of North and South America will show this pattern.

Considering the Mississippi River as a "major fracture," it is easy to recognize the "relieving fractures" as the rivers flowing into the Mississippi from the east and west; they do this at almost right angles and at almost regular intervals. The Red, Arkansas and Missouri rivers flow from the west; and, at about the same interval, the Yazoo,

Tennessee and Ohio rivers flow from the east. Where is the corresponding "crumpling?" Directly south of the Mississippi River in South America is the north-south Andes chain of mountains, the "major crumple," with relieving "minor crumples" of smaller mountain spur ranges running off to either side. According to Campbell, this is a fracture line dating from the *last crustal shift*; the one that shifted the pole from the middle of Hudson's Bay (where it was 17,000 years ago) to its present location which it reached about 10,000 years ago.

Campbell's theory explains something that he did not live to learn. Europe had four "recent" Ice Ages (Gunz, Mindel, Riss and Wurm), and North America also experienced four (Nebraskan, Kansan, Ilinoian and Iowan-Wisconsin). For many years most geologists were content to regard these names as just different European and American labels for the same phases of glacial advance. But very careful work on North American climates undertaken since the 1950's shows that these labels were not synchronous. Warmer phases in Europe often coincided with colder phases in North America, and vice versa.[30] Campbell's theory, as amplified after Campbell's death by Charles Hapgood of the University of New Hampshire, shows that these climate variations and differences can be neatly accounted for by Campbell's location of previous polar zones and the time taken for the crustal shift to be completed and stabilized (however briefly) in a new position.

We are now ready to take a look at the end of the Ice Age in Eurasia, the "Middle East" and Africa. The last crustal shift, according to Campbell/Hapgood, was from the middle of Hudson Bay to the present North Pole, and it left the Mississippi fracture and the Andes "crumple." But, according to Campbell/Hapgood, and according to any good atlas, one can easily see the fractures caused by the "second-last" crustal shift. Campbell said that the North Pole was located off the coast of Norway to the northwest during the worst of Wurm (Wurm I), and then it gradually moved with a crustal shift toward North America to end up in Hudson Bay.[31] If this line of shift is carried around the world in a great circular path it appears that it runs through the Middle East, through the Persian Gulf and out into the Arabian Sea. The floor of the Arabian Sea has a series of deep fractures showing that it was wrenched "recently."[32] Africa's famous Rift Valley may be evidence of the same crustal wrenching that took place along the weak seam between the African tectonic plate and the Indian Ocean plate. If we look at the Middle East in an "open-minded" way, while bearing Campbell's theory in mind, we can imagine, perhaps, that all of Africa was wrenched in a southwesterly direction. If we imagine this, we can suppose that the Persian Gulf, the Gulf of Oman and the Red Sea were once rivers. If Africa is mentally shoved back in a northeast direction, so that all these bodies of water around Arabia are closed up to river-like proportions, we notice another thing: that the mouth of the Indus river gets *very close* to the tip of the Arabian peninsula: Muskat.

Just one more thing has to be dealt with before we turn to *Genesis* in the Bible and the matter of Eden . . .

A good map of Egypt will show that the Nile makes a sudden half-loop at Qena, and that from Qena to the Red Sea is but a short distance, a distance traversed by the Wadi el Quseir. This is a (now) dry riverbed of formidable proportions; the modern road to Quseir follows its convenient route. It is interesting to speculate that, at one time in the not-so-distant past, the Nile River turned right at Qena to flow into the Red Sea at the modern town of Quseir... or, at least, to speculate that a considerable part of the Nile did so. Because, if it once did do this, a puzzling passage in *Genesis* suddenly makes sense (*Genesis* 2:10-14):

> And a river went out of Eden to water the garden; and from thence it was parted, and became into four heads.
>
> The name of the first is Pison: that is it which compasseth the whole land of Havilah, where there is gold.
>
> And the gold of that land is good: there is bdellium and onyx stone.
>
> And the name of the second river is Gihon: the same it is that compasseth the land of Ethiopia.
>
> And the name of the third river is Hiddekel: that is it which goeth toward the east of Assyria.
>
> And the fourth river is Euphrates.

If we look at a map, while preserving our mental picture of a "squeezed together" Middle East, we will see that the Nile would run, first, southeast but then be deflected northeast by the tall mountains of the Horn of Africa and that it would, indeed, encompass the land of Ethiopia. It would flow into the Arabian Sea somewhere along the coast of southern Arabia, perhaps not far from Eden. We can imagine the Tigris and Euphrates keeping separate identities down the length of today's Persian Gulf or, perhaps, one river split to run into the present lowland basin south of Muskat's mountains while one flowed through the present lowland Straits of Hormuz. The Indus River flowed in from the east. Now, if this was the case, and if, not so long ago the geography of the Middle East was "squeezed together" as I have suggested, then the four rivers of paradise actually existed once. Further—if we assume that one of the Tigris/Euphrates pair flowed south of the Muskat mountains—these rivers would have actually met in the shape of a cross; one Tigris/Euphrates entering the Arabian Sea from the north, through the present Strait of Hormuz; one Tigris/Euphrates entering from the west just south of the Muskat mountains; the Indus flowing in from the east; and the Nile flowing in from the south. These four "rivers of paradise" would probably have merged into one very great river, very wide but maybe not very long, that flowed into the Arabian Sea—a land with these mighty rivers that has now largely sunk to the fractured and wrenched bottom of the Arabian Sea. This place would have been the Garden of Eden, the home of humanity's earliest known "civilization." It may be that Eden was largely destroyed by the crustal splitting and wrenching that began the move of the North Pole from the coast of Norway to Hudson Bay. The cherubim and the flaming sword which guarded lost Eden may well have been volcanoes made active in the highlands all around.

Now, this has been an admittedly speculative and personal reconstruction of Eden. Is there anything at all to support it? The answer is a qualified "yes," a cautious "yes." If Eden was the location of a primal human culture (it might not, technically, have been a "civilization"), it stands to reason that survivors and refugees from Eden would have radiated away in all directions, but their obvious routes would have been the four rivers of paradise. It is now known that there was a very early civilization in the Persian

Gulf called "Dilmun," part of which was discovered by Danish archeologists on the Island of Bahrain.[33] Dilmun may be somewhat older than Sumeria, older than Egypt.[34] But, at the same time, it is known that Dilmun had some trade and cultural connection with the early Indus cultures; Dilmun weights and measures seem to have been related to, or identical with, Indus-culture weights and measures.[35]

If Eden refugees retreated along their river routes, there would be, and must be, close cultural parallels and affinities between Dilmun, Sumeria, the Indus cultures and Ancient Egypt—just as have been found with the unexpected discovery of Dilmun as the missing link. Perhaps all these riverine cultures were seeded by a people who had developed a primal riverine culture in Eden. Also, it is possible that the riverine cultures of Ancient Egypt and Sumeria (and Harappa/Mohenjo-Daro) did not take on their distinctive but interrelated cultural profiles not only because rivers offered easy agriculture, but because all of these early civilizations were daughter-cultures seeded from Eden with similar traditions and the same original language. Perhaps they named new settlements after old ones in Eden. Why else would the names of ancient places in Egypt, Sumeria and the Indus Valley be so similar, sometimes identical? There are really too many examples to name just a few. Any good, large-scale atlas that gives ancient historical sites easily proves this.

Aside from existing similarities in ancient Middle Eastern riverine cultures, there is another reason for speculating that at one time Africa was sort of "squeezed up" against Arabia (and Arabia squeezed up against modern Iran) and that the length of Africa extended in a more southeasterly direction into the Indian Ocean toward Indonesia. There was an enduring tradition, which lasted up until the 1480's in Europe, that, somehow, Africa curved eastward to meet up with the Spice Islands. Several medieval European maps show Africa this way; two are reproduced in this book (see pages 100-01). At least one faction of European geographers, admittedly a minority who turned out to be in error, believed that the Indian Ocean was an inland sea.

Why did this belief originate? Conventional historians maintain that this curious belief endured for 1700 years because of an observation by Alexander the Great's "Bematists" (scientists). On reaching the Indus River in 327 B.C. on their forced march from Bokhara, Alexander's army encountered crocodiles in the river. Knowing that crocodiles were found only in the Nile, Alexander's experts concluded that somehow the Nile and the Indus had to be connected. The only way that seemed possible was to have Africa curve around to the east, somewhere in the unknown southern hemisphere, so that the (unknown) sources of the Nile could hook up with the equally unknown source of the Indus. Crocodiles could then exist in both rivers. This incident with Alexander's army did happen, and the conclusions of the Bematists are well-known, but this alone does not fully account for the long-standing belief that Africa might curve eastward across the Indian Ocean.[36]

Also, we are faced with a puzzling ethnographic fact. The large island of Madagascar only 500 miles off Africa's east coast is not inhabited by Africans, as one might reasonably expect, but by Indonesians. How did they get there? If Africa has always maintained its present geographical orientation, these Indonesians can only have come by boat. It would have been the most audacious

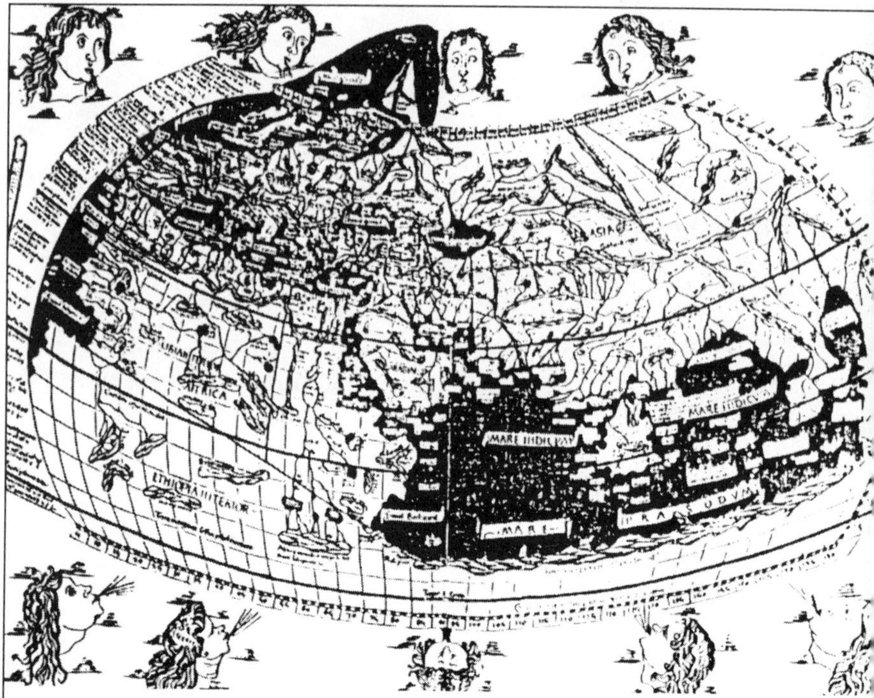

Johan Schnitzer's world map of about 1437 reflects the European notion that Africa curved toward the east, making the Indian Ocean an inland sea.

maritime mass-migration in history before European trans-atlantic voyages of the 16th century. The Indonesians had to cover 3500 miles of the Indian Ocean in ships, or boats, considerably less seaworthy and evolved than European vessels of some 2000 years later; and the Indonesians must have come in virtual fleets in order to account for Madagascar's population when Europeans first visited the place. Yet, in Indonesian history, folklore and tradition, there is not even a whisper of this great undertaking. Surely, if such a great maritime adventure had occurred, there would be some Indonesian record.[37]

However, if Africa did, at one time, extend considerably toward the east in the direction of Indonesia, then at

Another map from the 15th century which appeared in Hartman Schedel's *World Chronicle* of 1477. This map is based largely upon Ptolemy's geography of the 2nd century A.D. Note again that Africa is shown to connect with the Orient, leaving the Indian Ocean as an inland sea. The large island in this inland ocean is not Australia, but "Taprobane" (Ceylon, the modern Sri Lanka) and its much-exaggerated size reflects the immense wealth it was thought to possess, not its actual modest geographical extent.

least some of the problem is resolved. Indonesians could have reached Madagascar by island-hopping. There could have been a gradual migration, with nothing particularly noteworthy to record.

Campbell tried to give a rough chronology for his theory's crustal shifts, and this chronology is that the North Pole was northwest of Norway before 55,000 years ago; then it moved to the middle of Hudson Bay about 55,000-17,000 years ago; and finally shifted to its present location about 10,000 years ago or less.[38] By this chronology, Eden would have flourished with its four huge rivers while the North Pole was off the Norwegian coast and before it started to shift to Hudson Bay—that is, about 35,000 years ago at the latest. This is about seven times too long ago for the foundation of riverine cultures, according to conventional chronology! A 35,000 B.C. date for Eden is not acceptable!

However, Campbell's theory fits the Ice Age facts so well, and in detail, including the idea of massive releases of water as the old ice caps melt and before the new ones can absorb the run-off (possibly "The Flood") that it is very tempting to try to make Campbell's theory work!

The simplest way to do this is to propose that the crustal shift from the North Pole's position off the coast of Norway *to* Hudson Bay which happened sometime between 55,000 and 17,000 years ago—and let's pick a middle ground of 35,000 years ago which accords roughly with the Gottweig Inerstadial in Europe and the Iowan/Wisconsin transition (or "Interglacial") in North America—was the crustal wrenching or shift that created fractures *as rivers* (the Nile River flowing partly down the Red Sea and partly, perhaps, into the Mediterranean as today; the Tigris/Euphrates system of rivers flowing down today's Persian Gulf and into the Gulf of Oman, possibly by two different routes north (Straits of Hormuz) and south of the present-day Muskat mountains; the Indus river). These would have been *rivers*, albeit wide ones, and they would have flowed through a large tract of land, now sunken, that was in the present Arabian Sea. Africa would have been oriented, not north-south as is the case today, but "squeezed up" against Iran and India with its length trending southeasterly across the Indian Ocean to approach Indonesia.

Thus was Eden created by crustal movement, roughly 35,000 years ago; it was somewhere in the region now occupied by the Arabian Sea, possibly near Aden in present day South Yemen, unless that ancient name is no more than an irrelevant coincidence. It is worth trying to draw a mental picture of this lost world because only a few rather small geographic areas of the earth presently boast the same situation. It was a tropical to sub-tropical place in terms of latitude (with the North Pole then northwest of Norway) and, at sea level in the courses of the four mighty rivers running through it, it must have been a world of reeds (this is important), date palms, bananas, and many fish, both fresh and saltwater. Wild fowl, manatees, fresh water dolphins and (of course) crocodiles would have been plentiful.

This Eden boasted not only an ocean nearby (the Indian Ocean), but also numerous ranges of rather high mountains all around—the mountains of Muskat and Oman, those of now-dry Baluchistan, the mountains of the Horn of Africa, and the foothills of the Kithar Range of present-day Pakistan and India. Although Eden was tropical or sub-tropical at sea level, it could offer almost any desired climate, within a few miles from the rivers, *by altitude* instead of latitude.[39] As one climbed upward into the great highlands all around, different crops and different lifestyles must have flourished; tea, coffee and cotton must have been growable on these uplands; and it must have been possible to domesticate and herd native species of goats, sheep, cattle. Various kinds of cats must have tamed man and also breeds of dogs derived from the jackal.

This ancient Eden must have been a bountiful place where the species of plants, animals *and men* from Africa, Asia and Southern Eurasia met, mingled and contributed to a rich and varied economy within a great, but nonetheless limited, geographical region. In terms of climatic variation, today only Sri Lanka offers something similar but on a very limited geographical scale. But, in spite of its small size, this island was known as a land of wonder and magical romance to the medieval world (European, Asian and African). It was the "Land of Serendip" ("happiness") to Sindbad the Sailor and the fabulous land of "Tabrobane" to Europe. Its exaggerated size on many European medieval maps represents its supposed wealth and happiness, not its actual modest geographic extent. To some of the more ignorant rulers of the Middle Kingdom (ie. middle between heaven and earth, China) Sri Lanka (or "Ceylon") was the center of the world and they thought that "Rhum" (ie. "Rome") must have been there. The sophisticated Africans of the Zanj culture of East Africa thought that the Chinese porcelain they loved and imported in such great quantities (that East African coastal sites are now dated by the period of Chinese crockery now found in them) came from "Serendip," the land of happiness, and not from Canton (and the wily Arab ship-masters saw no reason to explain otherwise as this might lower the prices).

Aside from Sri Lanka, or "Ceylon," only one other place in the present world would have offered the climate of Eden by both latitude and altitude: the Caribbean coast of Colombia. But, this region lacks the biological diversity of Eden.[40]

Taiwan has sometimes been likened unto Sri Lanka, or "Ceylon," but its flora and fauna are too specifically Asian to match Eden's biological richness.[41]

Eden must have been altogether a remarkable place, a perfect "cradle" for the development of human civilization. And we may logically ask what type of people lived in Eden? It seems clear, from the excavations that have been done in this area (ie. around the fringes of the Arabian Sea) that three major groups, or "races," of humanity were there; first, and possibly foremost, among them are the "Capoids" (or "Bushmen") out of Africa; secondly, blacks or "Negroids" out of Africa; and thirdly, those almost ephemeral Fontechevade and Krapina people from Europe who had fled the Ice Age of Wurm I 35,000 years earlier. They, and a well-mixed mold of "them," almost must have formed the population of Eden. But there were others: if physical anthropologists are to be credited, then there was a significant admixture of "Australoids" in the Arabian peninsula at this time; the Indonesians (ie. "Mongoloids") must have migrated at least as far as Madagascar, although they may not have participated in Eden's culture.

This is not an attempt to concoct a conveniently "multiracial" mix in the admittedly conjectural Eden of this time (circa 35,000 years ago), rather this is what the current anthropological evidence indicates all around the fringes of this hypothetical Eden of this era; Capsians (thought to be Caucasians) throughout the Middle East;[42] Capoids (according to Carleton Coon) also throughout the Middle East[43] and evidence of Australoid presence in southern Arabia.[44] It is virtually certain that there must have been a large population of Blacks, or "Negroids," in Eden, too, although the earliest actual evidence of the Black major group of mankind comes from Asselar, in the Sudan, and is surprisingly recent—about 5000 B.C.[45], about the time that Eden perished or was a fond memory. However, this particular situation probably stems from lack of research, rather than an actual lack of evidence in this area. It is probably safe to say that Blacks were represented in Eden, and in large numbers. Ultimately, it was a racially mixed society. The only major group not represented in this Eden complex seems to be the Mongoloids; physical anthropology has not turned up any evidence of them around the Arabian peninsula. It may be that the later riverine cultures of China and Indochina owed little to what might be called the "Eden culture complex." Indonesians in Madagascar, if they truly came this early and not much later (by boat), seem not to have spread to mainland Africa. They remained isolated on their roomy island. This argument suggests that they did not come by boat around the time that the Christian era began, for surely if they had boats capable of crossing 3500 miles of Indian Ocean, they could have crossed the 500 mile wide Mozambique Channel to East Africa; they would have become a part of the Zanj culture—and an important part with their advanced maritime skills.[46]

If we want to conjure up a vision of the living culture of Eden, we probably cannot do better than read about the so-called "Marsh Arabs" of today.[47] They presently live in a shrinking area of marsh between the Tigris and Euphrates rivers. Their habitat remains much as the original Eden must have been (at least, near sea level), rich in fruits and vegetable food, with wild fowl and fish. Unfortunately, in 1992, their marshes cover less than about 6000 square miles of Iraq and their water is threatened with industrial pollution. Their lives depend heavily upon reeds, specifically the *berdi* reed. From this reed, they make their houses, clothing, cooking fires, fodder for their few domesticated water buffalo and all sorts of household utensils from sleeping mats to forks. Their islands are even artificially built up by an annual and purposeful addition of berdi reeds. Formerly, they made their boats of berdi reed, too, although now most Marsh Arabs have adopted a planked boat called a *masham*, of which there are three major types. According to the few

Europeans who have visited them, the Marsh Arabs lead an easy, tranquil but active life that contributes to remarkable longevity. A good, modern account is *The Tigris Expedition* by Thor Heyerdahl, a book that also provides color photographs of the Marsh Arab world. They probably live today much as the original "Edenites" lived many thousands of years ago, except that today's Marsh Arabs have adopted Islam (or, it was imposed on them), and their present society incorporates all the anti-feminisms, including purdah, characteristic of Islam in particular and of Judaism and Christianity in general. The original religion(s) of Eden were probably more African or Indian in character, probably pantheistic and possibly sun-worshipping, with a great deal more feminist representation in the "godhead" than modern Western religions incorporate. Certainly, there was no purdah in Eden! Even the Bible informs us that Adam and Eve went naked and unashamed in Eden, and the climate makes this probable. The inhabitants of Eden probably dressed much as the later Egyptians and Sumerians did: a short cotton kilt for both sexes, as ordinary wear.

Thor Heyerdahl went to visit the Marsh Arabs because he wanted to build the oldest type of ocean-going boat humanity had developed: the reed ship. This is important within the context of "The Flood." The usual type of large boat or ship in Eden almost must have been a reed craft, because reed boats are known to have been the very first Egyptian and Sumerian vessels. They are so ancient, and were revered so much, that even later ships made of wooden planks were often copied after the traditional reed boat design.

Heyerdahl made two previous reed ships (*Ra I* and *Ra II*, made in Africa of papyrus reed) and managed to cross the

Atlantic in *Ra II* in 1970.[48] By 1977-1978, Heyerdahl wanted to build a ship of berdi reed, a ship like the ancient Sumerians would have sailed 5000 years ago, a ship that was very probably the characteristic cargo carrier of Eden. His vessel was called the *Tigris* and he sailed it to Muskat, Oman, the Indus River in Pakistan, coasted Arabia and reached the Horn of Africa—it sailed over the region of Eden just the same way that similar reed ships had sailed many thousands of years ago.

"Noah's" Ark must have been a large reed sailing ship like Heyerdahl's *Tigris.* Biblical scholars accept the idea that the Hebrews borrowed the "Eden myth" and the "Flood Story" from the Sumerians, probably when Abraham and his people resided in the vicinity of Ur about 2000 B.C. The Sumerian Noah was called Ziasudra. In a later Assyrian text (but still a much older record than the Bible), it is said that Ziasudra built a huge ship *of reeds* in which to escape the Flood. The Dilmun hero, Utnapishtim, also built a reed ship *from the debris of his reed house* in which to escape the Flood.[49] Heyerdahl, in building three large reed ships, demonstrated their seaworthiness, stability, immense carrying capacity and one other quality: the fact that they could be built in virtually any size. The supply of reeds was the limiting factor and in Eden as it must have been, supply would not have been a problem. The Biblical "Noah's Ark" was supposed to have been 450 feet in length, which seems an exaggeration, but Heyerdahl actually saw a reed ship 112 feet long on the Tigris River.[50]

The Bible says that Noah's craft was made of "gopher" wood, and this word is often translated as acacia wood,[51] but it comes from much later Egyptian ship-building practices when acacia-wood planks were actually used *but only because the supply of papyrus reeds had become scarce.*[52] "Acacia" is actually *shittim* in Hebrew; "gopher" is a mystery.

With the story of Noah's Ark, we come to the destruction of Eden and the Flood—myths which, scholars agree, the Hebrews borrowed from the Sumerians of Ur. When did Eden die? Campbell places the North Pole "in" the region of Hudson Bay by about 17,000 years ago when it *began* to move to its present position. By 10,000 years ago, according to Campbell, the pole was stabilized in its present location. It is tempting to suggest that the reed world of Eden perished as the crust shifted to place the North Pole where it is today. The problem is that no one, not even Campbell, cared to guess how long a crustal shift might take: centuries, decades, months . . . or a few hours? Judging from the fact that many Siberian mammoths were killed instantly, for all practical purposes, and then transported very quickly into an arctic deep freeze from which they never thawed until their modern discovery (because their meat was still edible), a crustal shift *might* take just a few hours.[53] It might be a very sudden catastrophe—and we see that even the theory of the "Ice Ages" has evolved to support ridiculed Baron Cuvier.

Eden may have been destroyed within a few days about 8000 B.C.

It is tempting to suggest that this last crustal shift completed the work begun by the previous one: that it "opened up" the previous rivers to become the Red Sea, Persian Gulf and Gulf of Oman. Africa seems to have been wrenched in a southwesterly direction at this time. The Nile's course changed so that it flowed into the Mediterranean while Africa's Rift Valley may have been formed, or widened, in this same cataclysm.

And what about the Flood? It is an interesting circumstance that Campbell's theory can even account for this. Assuming that the North Pole was centered on Hudson Bay, we must bear in mind that this bay would represent but a small part of the arctic zone. It is surrounded on all sides by land and the bulk of the arctic ice cap would have been resting on Ontario and Quebec.[54] This means that this previous ice cap should have been much larger than the present arctic one, which is wholly on water. With the North Pole centered on Hudson's Bay, the *Antarctic* ice cap would still be situated over its present complex of islands.[55] In short, the antarctic ice cap might have been about the same size as presently, but the arctic ice cap should have been much larger.

This larger arctic ice cap would have contained much frozen water, perhaps several thousand *cubic miles* of water which the present, smaller, ice cap does not. This water would have melted with the change to the present situation, raising the sea level by many feet and flooding river basins. Studies indicate a rise in a sea level of about 300 feet between 8000 years B.C. and now.[56] It seems that much of the now-submerged "continental shelves" were above water within the time of cultured, "civilized" man. And, it is possible that most of the civilized human population likely lived near the sea shore or on the banks of navigable rivers simply because that made agriculture, trade and transport so much easier. Eden was not the only place drowned in the last crustal shift: many other human cultures must have perished also.

Does some of Eden still exist? "Yes"—*if this reconstruction of Eden is at all accurate*. One remnant bit of Eden may be Socotra Island off the Horn of Africa. Another surviving piece might be the Kathiawar Peninsula of India. Systematic archeology there might turn up some surprising evidence.

The Flood must have inundated a lot of inhabited land and killed off a fair proportion of coastal humanity. Campbell's theory of Ice Age mechanics, with the last North Pole largely on land, centered on Hudson Bay, argues that a lot of water was tied up in the ice of that ice cap. The Flood cannot have been some localized desert spate that impressed a few nomads. It must have been a genuine worldwide catastrophe. Indeed, if this reconstruction has any validity at all, the vast majority of evidence of early human civilizations must be underwater on the continental shelves or in now-desert remains of once-navigable rivers—such as in the interior of Oman, for example, where the real center of Dilmun may well be found. The island of Bahrain is only a small part of what Dilmun must have been.[57] We must bear in mind that not only the polar zones are re-arranged with a crustal shift, but all the other climatic zones changed, also. Thus, once-lush Eden is now in a desert zone for the most part.

Sumerians said that they came from Dilmun, originally, and that they came by reed ship.[58] Sumeria is supposed to have "started" about 4000-3500 B.C. and, like Egypt, it began with great and unexplained sophistication. Dilmun may have been a surviving example of the original Eden culture complex, or perhaps it was a daughter colony of refugees. But, given the carnage that must have accompanied the last crustal shift, perhaps 4000 years is a reasonable time to allow the human population to re-establish itself and to begin sending out new colonies; perhaps Dilmun,

Sumer and Egypt were those colonies and perhaps 4000 years of "Eden culture" are preserved somewhere in or around the Arabian Peninsula.

However, there was another "Eden," of a sort, in the region now occupied by the Sahara Desert. Manipulating a world globe according to Campbell's ideas will reveal that this Saharan area could not have been as lush as Eden because it would have been in a different climatic zone; but it would have been very habitable parkland with (satellite photos argue) many streams and rivers. Rock paintings at a place called Tassili, now almost dead-center of the Sahara Desert, depict people fishing, swimming and navigating reed boats. They also have horse-drawn chariots.[59] Again, they seem to be a mixed people; Blacks are definitely shown in the paintings, but also whites and (probably) "Bushmen." These whites seem to be the Fontechevade and Krapina people who fled from the Ice Age to sunny Africa.

This parkland died along with Eden. About 8000 B.C. it started to dry up because the last crustal shift had transported this region into another climate zone—a Northern Hemisphere desert zone. It seems logical that most of these parkland dwellers fled eastward until they encountered the Nile, an unfailing source of water. "Ancient Egypt" is popularly supposed to have "begun" when Menes unified Upper (the river itself above El Faiyum) and Lower Egypt (near the present delta at the Mediterranean). If this reconstruction is at all valid, could not this difference between "upper" and "lower" Egypt have been mainly cultural? Perhaps the inhabitants of upper Egypt were predominantly Eden refugees who had gradually expanded up the Red Sea/Nile, while the inhabitants of lower Egypt were predominantly refugees from the desiccated parkland

Figure 2:

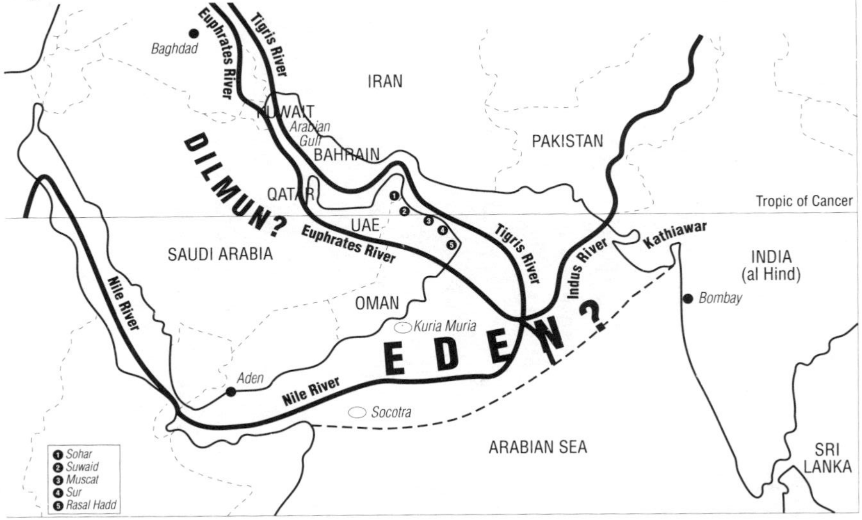

A Whimsical Reconstruction of Eden

I have assumed that the Nile once made a 180-degree turn at the site of present-day Qena to flow southward down the Red Sea. The Tigris and Euphrates were once separate rivers for a much longer distance and both flowed down the present Persian gulf; perhaps the Tigris flowed north of the Muskat mountains and the Euphrates flowed south of them. The Indus River continued its present course to complete the "four rivers of paradise". I have also assumed that the present seas around Arabia were once much more narrow than at present, the courses of broad rivers, and that Africa may have been directed significantly toward the southeast instead of due south.

to the west?

All over the world, the time "before the Flood" is remembered as a kind of "Golden Age" of relative plenty, relative peace—a paradise. This is not only the biblical position, it is repeated from China to Egypt among many cultures.[60] A lot of it may be nostalgia for a world forever lost, but there may be a grain of truth to the supposition that things were better "before the Flood."

The world certainly seems to have changed in a fundamental way. Whether by the mechanisms proposed by James Campbell or by some other agency, the world's climate zones did shift their location on earth. And this change of climate released a "new" kind of humanity into the realm of culture and civilization, a kind of man that had previously been more or less isolated within glacial Europe by his own special adaptations.

These were the "strongly Neanderthal" Caucasoids of Europe and Western Asia. They came out of their mountain fastnesses to conquer what was left of the civilized world, and they have been doing it ever since.

One early group to leave their mountainous domain were the Hebrews. They migrated southward from their Caucasus home and entered the civilized world of Sumer and Egypt. They were not the only Neanderthal-Caucasoid group to do this, but they seem to have been one of the earliest; and they left a history of their experiences. With these invasions, the world changed forever. It was truly the end of Eden.

IV
Images of a Glacial God
(Jewish Origins I)

Hebrew tribal history properly begins with a cultural hero named Noah, and it begins in the mountains of the southern Caucasus. The *Bible* does not say that Noah's ark landed *on* Mount Ararat, which is today a 16,946-foot peak just barely inside the eastern border of Turkey, but *in* "the mountains of Ararat" (*Genesis* 8:4). This could mean anywhere in the rugged geography of eastern Turkey, Armenian Azerbaijan, Armenian Georgia or Iran—in short, somewhere between the eastern end of the Black Sea and the western shore of the Caspian Sea where the southern Caucasus Mountains make a jigsaw puzzle of modern political boundaries. This area is a kind of "hub" or "nexus" of mountain chains: the Elburz range goes from it eastward across northern Iran toward the distant Hindu Kush and Pamirs; the Caucasus chain runs from it northward between the Black and Caspian seas to the steppes of southern Russia and the Ukraine; southward run the mountains of Kurdistan; while to the west the land gradually loses altitude to become the central plain of Anatolia.

This is the very place that is the center of late-lingering Neanderthal traits, a place, because of its ruggedness, that was avoided for a long time by post-glacial migrants from the south. This area was also avoided because formidable glaciers endured in these high mountains for a long time after the end of the actual Ice Age, the very place where folklore about *kaptars* concentrates and where living specimens have twice been reported by Soviet zoological and medical experts within the past 51 years: this is the place where the Soviets targeted their most massive of four expeditions to find the "snowman" of the U.S.S.R. It is a place surrounded by sites where Neanderthal skeletal remains have been uncovered by modern anthropologists. It was a major Neanderthal refuge.

Scholars accept that the"Flood Myth" as related in *Genesis* 7 is borrowed almost verbatim from various Sumerian and Akkadian accounts that the Hebrews heard when they reached the neighborhood of Ur about 2000 B.C. Scholars also accept that the "Garden of Eden" myth is also a borrowing from Mesopotamia.[1] Yet, it is obvious, too, that genuine, if vague, tribal memories have been woven into these foreign, borrowed stories. The flood tale is seamlessly joined to the person of Noah, who may have been a real historical character (or whose name may be that of an early tribal dynasty of chiefs) since a genealogy of sorts leads from him to Abraham (an undoubtedly historical individual). If the Flood was really a worldwide catastrophe of major scope, even the proto-Hebrews in their mountains probably experienced some of its effects, although not so drastically as coastal peoples. *Genesis* 8:13 informs us that "the face of the ground was dry" on the first day of the first month. This seems a little convenient. What more probably happened was that after the flood disrupted the continuity of whatever culture had previously existed, at some point the proto-Hebrews regained sufficient cultural continuity to begin a new reckoning: the first day of the first month; they started a new calendar. And it is this reckoning "after the Flood" which persists to our own time, although some of the very early chronology is sketchy. Our world is the one that began "after the Flood." Only hazy and allegorical memories remained of the lost world "before the Flood." Eden is one such memory, but one borrowed by the later Hebrews from people nearer the coast of the Persian Gulf.

Yet, woven deftly, although very sketchily, into the opening chapters of *Genesis* are some Hebrew tribal memories that were not borrowed. These passages have puzzled Biblical scholars and everyone else.

The story of Cain and Abel is a bit non sequitur. Cain was an agriculturalist. Abel was a herdsman (and, presumably, part-time hunter). The tale is usually interpreted as a conflict between pastoral herdsmen and agriculturalists. But it is interesting that, in these mountains, the herdsmen must have been native to the area (or at least established there longer than agriculturalists), while the agriculturalists could only be incoming migrants from elsewhere, probably Mesopotamia—where agriculture had been invented "before the Flood" (if the extensive commerce of Dilmun is any guide). The story informs us that although the agriculturalists finally "won," at first, at least, the pastoralists were considered to have greater prestige. *Genesis* 4:3-5 . . .

> And in the process of time it came to pass that Cain brought the fruit of the ground (as) an offering unto the Lord. And Abel, he also brought the firstlings of his flock and the fat thereof. And the Lord had respect unto Abel and his offering. But unto Cain and his offering he had not respect. And Cain was very wroth, and his countenance fell.

If we are to read anything at all into this story, it may be the memory of a primal conflict of life style between the resident mountain herdsmen and incoming migrants who brought agriculture. It must date from a time in the very distant past when the resident herdsmen were dominant and more numerous. God was "theirs" and accordingly endorsed their life style. Nonetheless, Cain killed Abel, and the agriculturalists became dominant in the tribal economy; yet the prestige of the herdsman endured as a memory. If it means anything, this story of Cain and Abel recalls an uneasy fusion of two separate types of people.

Another passage seems to reflect the same thing. *Genesis* 6:4 . . .

> There were giants in the earth in those days; and also after that, when the sons of God came in unto the daughters of men, and they bore children to them, the same became mighty men which were of old, men of renown.

This is a familiar passage from the authorized King James version. It is familiar because of its puzzling nature (once heard, it cannot be forgotten), but also because many recent writers on the "ancient astronaut" theme from von Daniken onward (and backward)[2] have used it to infer that celestial beings must have come down to interfere, genetically, with earthlings. This may seem unlikely, but the passage itself is sufficiently inexplicable to spawn any number of bold speculations. Christians believe as an article of faith that God had only one son in any strictly parental sense. Jews and Moslems deny even this. Who, then, are these "sons" of God? Why are they mentioned in an early phase of Jewish history?

There may be a most interesting answer about the identity of these "sons" of God and an equally interesting one pertaining to their importance in Hebrew history. Remember, they fathered children who "became mighty men of old, men of renown," who must have been chieftains or heroes, or both of later Hebrew history because these giants existed not only "in those days," but "also after that."

The answers depend upon the fact that this passage, so familiar, is actually a serious mistranslation from the original Hebrew. Yonah N. Ibn Ahron, a Jewish rabbi, is not only intimately familiar with the *Old Testament*, but an expert on the vast body of Hebrew exegetical work: written commentaries on Hebrew religion, life and legends preserved by a long and illustrious line of rabbis from the earliest times. Although these exegetic books are virtually unknown to the great majority of non-Jewish people, they actually comprise many times more history than what gentiles know as the *Old Testament*; they are really the basis for rabbinical study of the Jewish scriptures.

Rabbi Ibn Ahron is, however, an expert in two additional respects which make him absolutely unique in the world of scholarship: in addition to being a rabbi, he is a professional linguist (speaking over twenty Middle Eastern and Eurasian languages). He is also extremely well-versed in human and "pre-human" anthropology and paleontology by virtue of his studies under Cambridge zoologist, Dr. Ivan T. Sanderson. In an extract written by Rabbi Ibn Ahron, references to exegetical books are parenthesized, sometimes citing the title of the book, but sometimes the author of it. He also parenthesizes Hebrew phrases where necessary to clarify an interpretation. Rabbi Ibn Ahron is quoted here *verbatim*, brackets and all, with no attempt to edit or simplify his style.

He says of these "sons of God":

> The earliest Biblical reference to genetic variation within the human family is in the sixth chapter of *Genesis*, where we read: "And it was when Man began to multiply on the face of the earth and daughters were born unto them and the Sons of Those from on High (Jerusalem text: those who are worshipped) saw the daughters of Man, for they were good-looking, and took unto themselves wives from such as they might care to choose. And Yehova said, "My power can never benefit the perverted ones who have made themselves human; and their days shall (number) one hundred twenty years. In those days there were (already) nephilim in the land; it was thereafter that the Sons of Those from on High came unto the daughers of Man, and there was born to them those Giborem who were ever after considered to be a divine people (lit.—the people of the Name)."
>
> Nephilim is often translated "giants," but the commentators tell us that they were so called because men would fall (nophel) on their faces with fright at the sight of them. (cf. Ibn Ezra, B.K.) The giborim, who are later referred to as giborei tsayid, are reputed to

> have been "as tall as a tree" in contrast to the shambling nephilim (cf. Yoma). Giborei tsayid means "The Mighty Ones of the Hunt."[3]

If we can rely on this as a scholarly and accurate translation, with knowledgeable commentary, from the original Hebrew, then some very provocative possibilities present themselves. First, we can dispense with these "sons of God." They are a figment of bad translation. We are left with the "sons of those from on high." These people, whoever they were, considered the women "of man" to be good-looking—and presumably this means *better looking than their own females*—because they would come (down?) to take "unto themselves wives from such as they might care to choose." In short, those from the heights would come down and rape (in the original sense of "carry off," with forced intercourse an inevitable consequence) the women "of men." Without resorting to either mystical or extraterrestrial conjectures, we can say, at least, that those from on high and "men" presumably down below *were of the same species.* Women bore children to them. The two groups were "interfertile," the same species. It is equally certain, however, that "those from on high," on the one hand, and "men" on the other hand, *were of markedly different races.* They are strongly differentiated in the text.

Now, it is not made absolutely clear, but it seems to be the case that these people from on high are the same as the nephilim—which is to say, human beings so different from "men" that men would fall on their faces with fright at the sight of them, and in tradition (cf. Yoma) these nephilim *shambled*—by which we may infer a heavily built, somewhat awkwardly articulated human being whose appearance was frightening to "men" (below). Given that this tale comes from a time "before the Flood" when the proto-Hebrew tribes were in the Caucasus Mountain region, the situation seems fairly obvious. This is a tribal memory of a very ancient time when "more modern men" trickled into this montane backwater and encountered resident Neanderthals. The incoming migrating peoples naturally avoided rugged mountains as far as possible, if only because they were agriculturalists and crops require somewhat level terrain, and stuck to plateaus and valleys as routes of penetration. Once in the mountains, they began to encounter "those from on high": Neanderthals from the higher valleys and still-glaciated alpine regions of the true uplands. These Neanderthals were certainly hunters, and they may have even evolved into herdsmen by then (roughly 8,000-10,000 B.C.?). In any case they preferred the "more modern" women to their own. They raided the incoming scattered groups of people and carried off their women. From this pattern of enforced unions, a new and composite people emerged: the giborem tsayid, the "mighty ones of the hunt," mighty men of old, men of renown . . . in short, *the ancestors of the later Hebrews.*

These giborem were much taller than the shambling nephilim who presumably fathered them, and they were markedly taller than even the more modern men whose women they raped. It is *possible* that these more modern incoming humans, the "men," were of slight build and relatively short stature: they must have come from the south, and may have been of predominantly Capoid genetic stock. Capoids were never very large people, although they seem to have been a bit taller in the past than today's "Bushmen" who are barely outside the size range of known modern pygmies.[4] In any event, all the "Edenites" tended to be fine-boned and of only moderate stature, as Sumerian remains attest (and possibly, again, because of significant Capoid genetic input).[5]

World-famous anthropologist, Dr. Harold L. Shapiro of Columbia University, has long been interested in the phenomenon of stature increase when ethnic or racial groups are allowed to intermix. During the American Revolutionary War, for example, the British considered the Americans to be giants. A comparative study of skeletons shows, in fact, that the Americans were significantly taller, on average, than their British adversaries. This phenomenon has always been chalked up to better nutrition, but Shapiro suggested another explanation[6]: genetic mixing. Most of the Americans of 1776 had originally come, generations before, from the British Isles and yet within a couple of centuries (at most) had grown visibly larger. Shapiro suggested that this was because, in the British Isles themselves, a man from Herefordshire (for example) might seldom or never get the chance to meet with and marry a woman from, say, Scotland. Until well into the 19th century, most gene pools in the British Isles remained isolated from each other. People married within a fairly small radius. And it is well known that genetic isolation leads to diminishing stature.

However, the situation changed in America. People from all over the British Isles emigrated to America. Here, unlike at home, they were thrown into a true melting pot with people from all over Britain (and elsewhere) living side-by-side in New York City, Boston and numerous smaller towns and villages. For the first time, British genes (and others), isolated to a significant degree at home, had the chance to mix. That, Shapiro believes, and not nutrition, accounted for the very rapid increase in size among Americans. A similar phenomenon is visible now in Japan. Before the 1850s, Japan was a feudal society. Even up until the 1950s there remained a very real degree of genetic isolation because transport between the islands comprising Japan was limited. Most of the population married people from within a radius of just a few miles. But because of World War II, and after it, young Japanese were thrown together in the major cities and formerly isolated genes could mix for the first time. The result is plain to see in any modern street scene of Tokyo: younger people are very much taller than their elders.

All of this has relevance to the giborem tsayid. They, too, were the result of a mixture between formerly islolated gene pools. The isolation of the Neanderthals in the highlands needs no further exposition, but even the incoming groups of migrants from the south must have travelled in relatively small and genetically isolated clans or tribes. It should not be surprising that mixing between them and the local Neanderthals would have produced noticeably taller offspring "as tall as a tree" (as Japanese fathers often say to towering sons and daughters today).

In Rabbi Ibn Ahron's re-translated and annotated passage from *Genesis*, he suggests something more provocative and more profound than merely a highly Neanderthaloid ancestry of Hebrews. If Ibn Ahron's words are very carefully read, a subtle confusion in orientation and perspective is evident. And, we must bear in mind that his tale of the giborem was a very ancient Jewish tradition that was only *written down* much, much later. It *seems*, from this retranslated passage, that Yehova, or "God," must have been the *God of the Neanderthals* and not the God of the "men" involved. It is easy to see how slightly-built and outnumbered "modern men" might readily appease (worship) numerous and frightful-appearing Neanderthal adversaries against whom they were virtually powerless. It is possible to understand how they allowed their women to be taken away by the *highlanders*. In this passage, Yehova seems to chastise the highlanders for wanting to become "human," and elsewhere the offspring of the highlanders and the luckless "daughters of men" are considered divine, to be people of the Name (ie. knowing the name of God, Yehova). The implication is clear that in this dim proto-Hebrew past "before the Flood," the power and "God's approval" lay with the ones from on high.

This is precisely what we saw in the story of Cain and Abel; both tales may date from roughly the same age (8,000-10-000 B.C.?). Abel, the herdsman, finds favor in the sight of God, while Cain, the agriculturalist, is repudiated.

In this history we witness a gradual evolution of perspective. First, the ones from on high, our putative Neanderthals, are militarily and socially dominant over the "men." God belongs to the dominant Neanderthals, and God is obviously biased toward them—or, maybe it would be better to say that since the Neanderthals were dominant, their god was also dominant. The offspring of those from on high and the hapless "human" females are considered to be divine, possessing the name of God. But even Yehova complains that he cannot do much for those who will pervert themselves into becoming "human."

Then, later in the *Bible*, as the two races become more completely melded and mixed into one, the distinction between those on high and "men" is gradually dropped.

Everyone becomes "men," the original Neanderthal God becomes the God of "men," and agriculturalists become equal in prestige to pastoral herdsmen (even though tradition preserves the former greater status of the herdsmen). In short, in the *Bible* we witness a gradual change in perspective from Neanderthal dominance, and the dominance of *their* God, to an evolving situation of a composite population claiming the originally Neanderthal God.

It has too often been stated that the Neanderthals "died out" under the onslaught of more modern people. This notion is, and has been, simply an expression of European wishful thinking because, again, we much preferred to consider artful Cro-Magnon as our racial ancestor rather than artless Neanderthal.

Now, in certain places in Europe, it is probable that the Neanderthals were largely exterminated by incoming migrants. These places would be along the path of the major post-glacial migration routes where incoming tribes could reasonably be expected to have outnumbered the resident Neanderthals considerably.[7] Along such routes, there could not have been much genetic mixing with Neanderthals. The later population exhibited very weak Neanderthal traits and the remnant Neanderthals retreated (if they could) into mountains where the present population shows somewhat stronger Neanderthal traits—but the pure Neanderthals were killed off, to be remembered as trolls and gnomes of European folklore.

In other places, though, the opposite happened. In regions that were backwaters from major post-glacial migration into Eurasia, where "more modern" migrants came much later and in scattered clans, not hordes, the resident Neanderthals must have had the upper hand for a long time. Neanderthal power, genes and beliefs (even language?)

must have initially prevailed over the outnumbered newcomers. Neanderthal physical traits would have become diluted only very gradually over many, many generations. This re-translated passage from *Genesis* describes just such a situation, and in just the place where it is almost certain to have occurred. A place where nearby Neanderthal anthropological sites, plus the physical characteristics of the modern population (plus the presence of *kaptars*) proves that things happened this way.

We are left to conclude that the original Hebrews were a very strongly Neanderthal population and that their God was a largely Neanderthal conception.

According to the *Bible*, if I have counted the "begats" correctly, there were twelve generations from Adam up to Noah and the Flood. This is probably allegorical. As we all know, the life spans "before the flood" were improbably long. Some of these names may more likely be the names or titles of tribal dynasties. Again, if one has counted the "begats" correctly, there were eleven generations from Noah to Abraham. These "generations" may be the duration of tribal dynasties.

Since we cannot really trust the chronology "before the Flood," although the major post-glacial realities seem to be remembered more or less accurately (in an allegorical way) through a combination of Hebrew memory and borrowed Sumerian memory, the tangible history of the Hebrews must begin with Noah. In his time, or a bit later, the Hebrews began a migration toward Mesopotamia. They reached the city of Ur in the time of Terah, Abraham's "father" (or tribal chieftain) and lived near Ur during part of Abraham's life. This was about 2000 B.C., according to most Biblical scholars.[8]

It is about 630 airline miles between the actual Mount Ararat and the site of Ur, so this early Hebrew trek from their natal mountains must have taken some considerable time. They did not march purposefully toward Ur and Mesopotamia. It is unlikely that they had any planned destination in mind to begin with. They moved as nomadic herdsmen and primitive farmers, from year to year as grazing and sowing seemed best. Emerging from the mountains of Kurdistan about 300 miles from Ur, but close to some of the outpost towns of northern Mesopotamia (called Akkadia) they would have learned of civilization and seen some of its works. This may have hastened them more purposefully southward.

In this situation, the Hebrews are coming from the "north," but this is meant in a very general sense because no one knows where they started their migration. *Genesis* 11:2 says:

> And it came to pass, as they journeyed from the east, they found a plain in the land of Shinar; and they dwelt there.

This plain of Shinar is accepted as being Mesopotamia in the vicinity of Ur, which is to say western Mesopotamia along the Euphrates river. The biblical statement about travelling "from the east" may refer only to the last leg of their migration toward the specific city of Ur in Sumeria. But if it truly reflects the general trend of their migration out of the mountains, then their original homeland must be placed further east than the Caucasus mountains proper, the Elburz range at the southern end of the Caspian Sea, perhaps: still a Neanderthal refuge.

When they reached Ur in Sumeria, the city was already

about two thousand years old. Twentieth century excavations, notably by Sir Leonard Woolley in 1929, demonstrate that Ur was founded about 4000 B.C.[9] However, Ur may have existed for an unknown length of time "before the Flood" because Woolley found a layer of homogeneous mud 10-13 feet thick below the city's debris and, beneath the mud, confused evidence of earlier occupation.[10] If we consider the mud as detritus from the Deluge, then Ur may have existed before it. Among the 35,000 clay tablets found in the Sumerian ziggurat of Nippur,[11] what appears to be the original "Flood story" turned up: survivors on a big ship *of reeds*, after being swept for some days on turbulent waters, landed first at Dilmun and then sailed on to Ur.[12] But inference is clear in this story that the survivors on the reed ship recognized Dilmun and, having arrived there and getting their bearings, sailed purposefully for Ur. Therefore, the implication is that Ur had existed before the flood, but was wholly destroyed, for all practical purposes, and had to be rebuilt.

In short, Ur was an ancient city when the Hebrews arrived, and its civilization was rich and varied. Sumerian culture must have impressed these primitive nomads, and it may be that it was easier to make a living on the fringes of a great civilization than in the mountains from which they had come. They lingered around Ur for some time—the Bible is not very specific on this point—but they subsisted on the fringe of Sumerian civilization long enough to adopt a Semitic language.

Jews today label their critics as being "anti-Semitic," as if the term refers to an ethnic group or (loosely) a "race," and claim thereby a Semitic origin. But the term *Semitic*, properly speaking, refers only to a group of languages, not an ethnic, cultural or racial group.[13] The Semitic languages are supposed to have originated in northern Arabia,[14] according to most authorities, and the living "type language" of that linguistic family is Arabic. However, common sense alone argues that any group of people related closely enough to evolve a recognizable language must have some cultural and genetic affiliation. Therefore, perhaps, we can imagine "Semites" as having been an early mix of Capoid, Capsian, Negroid and (possibly) Australoid people who met and mingled in northern Arabia between 35,000 and about 10,000 years ago—they are, in fact, our "Edenites." If they truly came, *or their language came*, from northern Arabia, then this places them in the now-desert region where Dilmun once flourished, a part of the "Eden culture complex" or a daughter-colony of Eden, of which the island of Bahrain (about 250 miles southeast) is the only known remaining bit. In short, the original Semitic-speaking people came into Mesopotamia from the west and south, the exact opposite direction from which the Hebrews entered Mesopotamia (the north and northeast).

The Biblical Hebrews were not Semitic at all in any cultural or ethnic way. They merely *adopted a Semitic language*, one perhaps akin to some now-vanished dialect of Akkadian-Sumerican spoken in the immediate vicinity of Ur. Therefore, when Jews imply a claim to a Semitic "racial" and cultural heritage by labelling their critics "anti-Semitic," they are abusing the evidence and simultaneously denying their own Caucasus roots.

Hebrew tribes may have adopted a Semitic language while residing in Ur, but the most fundamental aspect of the Hebrew language betrays an ancient non-Semitic origin. This aspect is the letter-names of the Hebrew language.

Figure 3:

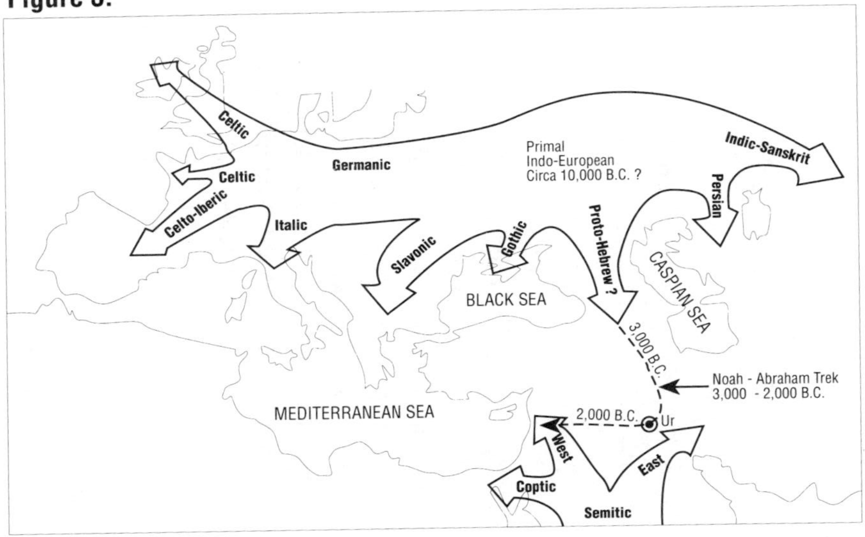

Major Caucasian Linguistic Groups

This map is based on many modern sources. On this evidence, and that of the *Bible* itself, the early Hebrews must have spoken an Indo-European language at the legendary time of Noah. They adopted a eastern Semitic language while in the neighborhood of Ur in Abraham's time about 2000 B.C., and joined the western Semitic language group as they dwelt in Palestine.

Table of Letter-Names

Letter	Greek Name	Hebrew Name	Irish Name	Species of Tree
A	alpha	aelph	alev	silver birch
I	iota	jod "yi-od"	ioda	yew
R	rho	resh	ruis	elder
B	beta	beth	beth	birch
N	nu	nun	nion	ash
E	eta	heth	edha "etha"	white poplar
M	mu	mim	muin	vine
O	o "ah"	ain	onn	furze
L	lamda	lamed	luis	rowan
D	delta	daleth	duir	oak
G	gamma	gimmel	gort	ivy
T	tau	tav	tinne	holly
S	sigma	samech	saille	willow
Z	zeta	tzaddi	straif	thorn
K	kappa	koph	quert	apple

The acclaimed English linguist, poet, author and cultural historian, Robert Graves, presented the following comparative table in two divisions (see table of letter-names).[15] In the upper lists, the Hebrew letter names most closely resemble the letter-names of the Irish branch of Gaelic. In the lower table, the Hebrew letter-names most closely resemble ancient Greek letter-names. The difficulty here is that both Greek and Irish are well-esteemed members of the Indo-European family of languages, not the Semitic family of languages. Hebrew letter-names should not resemble Greek and Irish ones at all—and yet, they do! Greek and Irish diverged from their parent Indo-European language a long time ago. No one knows exactly how long ago, but perhaps 5,000-7,000 B.C. might be a good guess. The similarities of Hebrew with both (in some ways) indicates that the Hebrew letter-names date from a time when Greek, Irish and Hebrew had not diverged so far from the parent Indo-European tongue—in short, earlier than, say, 7000 B.C. At that time, the parent Indo-European letter-names were similar, but not quite identical to *any* of the later-diverged Greek, Irish or Hebrew letter-names. This means that the Hebrew language betrays an ancient Indo-European origin (although in biblical times and more recently it has become almost wholly Semitic).

Now, the Indo-European language is thought to have originated either in the Caucasus Mountains or in the Russian-Ukrainian steppes just to the north.[16] Naturally, no one is certain because this ancient language (now vanished in its original form, but confidently thought to have been basically reconstructed over the past 150 years by linguists)[17] was probably spoken for the last time about 10,000 to 12,000 years ago.[18] Since then, it has evolved into the various Eurasian tongues spoken from the Pamirs and India in the east (Sanskrit) to Ireland in the farthest west (Irish/Goidelic). Perhaps the Basque language of the Pyrenees might be a surviving example of a language spoken once, long ago, by Neanderthals of the region. But Basque may be an example of the language spoken by some of the first Fontechevade-Krapina people returning to Europe during the Gottweig Interstadial. It is true that the Pyrenees are mountains, and it is also true that Neanderthals did live in them—but it also happens that these rather modest mountains are directly in the path of the major migration route into post-glacial Europe: the Iberian Peninsula. It may be that in spite of their modest montane haven, Neanderthals in the Pyrenees were totally inundated by newcomers, leaving only a vestige of Neanderthal physical traits in the sturdy

Basque people and (maybe) a composite language of Neanderthal and Fontechevade-Krapina speech. It is unlikely that anyone can ever clarify this matter.

However, since Indo-European languages are known to have originated in what is known to have been a "genetic refuge" for late-lingering Neanderthals, and since this area was remote from major post-glacial migration routes, it is tempting to suppose that primal Indo-European may well retain significant elements of Neanderthal speech. It would be interesting to know if there are any close affinities between the supposedly primal (reconstructed) Indo-European and the modern Basque.

The geographical location of "Old Indo-European" can be pinpointed with some accuracy. Naturally, linguists have traced Eurasian back to this Caucasus-steppe region, but botany really provides the definitive evidence. It happens that the letter-names in all Indo-European languages were once, long ago, the names for trees. Most modern Eurasian tongues have evolved and changed so much that this is no longer very apparent. But a language like Irish (the Goidelic branch of Celtic/"Gaelic") migrated quite early out of the Caucasus heartland, encountered few foreign linguistic influences as speakers of it travelled through a largely forested and unpopulated Europe, and then found itself more or less isolated in Ireland. Old Irish letter-names were still the names of trees and, since this language was spoken up until the medieval period and numerous literary manuscripts in it still survive, the trees corresponding to the letters are well known. Irish poets even called the evolution of their language and alphabet "The Battle of the Trees" as Roman, Anglo-Saxon and Germanic linguistic elements intruded upon their Celtic preserve. *Cad Goddeu* ("The Battle of the Trees"), a humorous poem by the celebrated 12th century Irish bard, Taliesin, exists in several translations; it lists the trees of the alphabet.[19]

Therefore, to find out where this very ancient derivative tongue of primal Indo-European originated, all one must do is find a place where *all* of these trees grew together in Eurasia. There is only one such place: the Caucasus region between the Black and the Caspian Seas. The letter-names of Old Irish, like the letter-names of Hebrew itself, preserved the names of Caucasus trees even though some of the trees did not grow where Jews and Celtic Goidels finally ended up. The apple provides an excellent example. It is thought to have originated in the Caucasus at the eastern end of the Black Sea. Groves of what are thought to be primal, wild apples are found only there.[20] Apples did not reach Ireland until early medieval times, but the Goidelic language had long had a name for it. This fruit is mentioned in ancient, traditional stories, even though it did not exist on the island. Jews present an identical parallel. Apples did not grow in Palestine in Biblical times, and it has only been since the 1930s that marketable varieties have been developed to grow in the Levant.[21] Nonetheless, the Hebrew letter-names preserve the ancient Indo-European name of this tree, and Biblical literature even refers *allegorically* to apples (the one of Garden of Eden fame springs to mind) when they did not exist in the Hebrews' new home. When the word "apple" is *botanically* used in the Old Testament, scholars think that either the pomegranate or the apricot is intended;[22] the Hebrews had to substitute.

Before leaving the subject of apples, and since the Garden of Eden has been mentioned again, we should perhaps enquire into this "fruit of the knowledge of good and evil." The *Bible* never says that this fruit was an apple. And, if the story was borrowed from the Sumerians, there is a good reason why apples are not mentioned as the "fruit of knowledge": the Sumerians had no apples. Yet, apples have always been connected with Eve and the end of Eden. Why should this be? Why should Hebrew tradition preserve such a detail if Jews originated in a Semitic language region which had no apples?

The reason is the "K" or "Qu" sound, the letter-name for the apple. In primal Indo-European, as painstakingly reconstructed by five generations of linguists, the "kwe" root signifies questions. Even our modern word *question* itself preserves this initial ancient Indo-European root-sound. In Latin, another Indo-European derivative language, almost all "questioning words" begin with *quo*, the old Indo-European root-sound *kwo*. We have, for instance, the question "Quo vadis?" ("Where are you going?"). In the old language of Iran/Persia, the "question sound" was *ku*. In Old Irish it was *cw*. In all Indo-European derived languages, the letter-name which is also the apple, is the sound for asking a *qu*estion. Anyone who asks questions is likely to acquire a certain amount of knowledge. The apple thus symbolizes knowledge, which is why we polish one for a teacher (in hope of acquiring some of his/her knowledge), why it was the fruit of the "tree of the knowledge of good and evil" in the Garden of Eden, why ignorance is bliss—and why Eve should have left it alone!

It is most interesting that this "tree of knowledge" was in the middle of God's garden. *Genesis* 3:3...

But of the fruit of the tree which is in the midst of the

> garden, God hath said, Ye shall not eat of it, neither shall ye touch it, lest ye die.

King Solomon, a wise man, had seven pillars of wisdom which supposedly "supported" the temple. Now, it is well known that within this temple there was a grove of trees, seven trees, and the middle one was an *allegorical* apple tree (actually, either an apricot or pomegranate substitute).[23] One might say that the Jewish "Temple" or "spiritual construct" was *knowledge of God*, which is why the apple tree was the central support of it in a symbolic sense. This is also why the menorah, the Jewish candlestick with seven stems, often has the longest, central stem decorated with pomegranates. It is a central mini-pillar symbolizing the need for knowledge of God.

These ancient and intricate inter-relations of primal Indo-European language and symbolism could be carried on almost ad infinitum—as Robert Graves did in *The White Goddess* and as Sir James Frazer did in *The Golden Bough*—but this has been more than enough to demonstrate the actual Caucasus/Indo-European origin of the very ancient Hebrews.

If the ancient Hebrews were truly of strongly Neanderthal genetic stock, then surely the *Old Testament*, in addition to the vast body of Jewish exegetical writing, would have a lot more to say about it. And indeed it does...

Rabbi Yonah N. Ibn Ahron has the following to say as he continues his discussion of genetic variation within the Israelite tribes:

> Our main concern must, however, rest with the creatures who terrified the Israelites during the

Exodus from Egypt and their period of wandering in the Sinai desert. These were the Sheidim—the Destroyers—who had been known to the Patriachs (Abraham, Issac and Jacob) as the Seirim—the Hairy Ones.

The best scriptual description of their characteristics may be drawn by inference from the account given in *Genesis* 27 of the manner in which Rebecca contrived to win the Rights of the First-born for her youngest son Jacob, as against the prerogatives of his twin brother Esau, who is described as "coming forth first from the womb 'reddish with a great mat of hair,'" (as thick as a wig as Rashi puts it). Esau grew up as a hunter, very much ashamed of his deep red covering, which earned him the nick-name Edom, or Reddy, as the vernacular might have it. The implication of the text at every turn is that, although ostensibly Jacob's twin brother, Esau was no true son of Rebecca. It would appear, at least that she felt that way about it, because she forced Jacob to seek the blessing of his father, Isaac, whose weak sight made possible the success of the ruse. To this end, Rebecca slew two young goat-kids (lambs, according to some sources), and clad Jacob, whom the text describes as her "smaller" (and not "younger", as one would expect) son, across the back of his neck and around his wrists in their hide (*Genesis* 27:19 ff.) Thus prepared, Jacob went into the presence of Isaac, his father. It is from this interview that we learn the most about the Hairy Ones, insofar as specifically Judaic sources are concerned.

Jacob was successful in passing himself off for Esau on three counts: (1) his apparent hairiness, which we have just accounted for; (2) his voice, and (3) the odor of his clothing. Jacob's speech, when he entered his father's chamber and identified himself as Esau, is thought by many not fully to have convinced the wily

> Isaac, but this impression is based on a misreading of the text. Isaac is quoted as saying "the hands are the hands of Esau, but the voice is the voice of Jacob." In reality, the passage reads, "the hands are the hands of Esau and the voice of Jacob is higher pitched (than that of the one who stands before me)"—kal kol (as the passage is actually written) instead of kol kol, the popularly accepted reconstruction. Indeed, the commentators bear out this interpretation when they tell us that Jacob counterfeited the voice of Esau by growling from deep within his throat (cf. Klei yakar, etc.). The outdoorsman odor of his clothing was achieved through more obvious means, for Rebecca simply appropriated the cast off garments left over from Esau's last visit. The stench, we are told, was akin to "a field of rotten potatoes."[24]

Jacob was a direct descendant of Abraham, a succeeding patriarch of the Hebrew tribe(s), and so this incident of deceiving Isaac in order to gain Esau's legally rightful inheritance cannot have taken place much later than about 1900 B.C. Yet, about six hundred years later, the Hairy Ones bedeviled the Israelites during their Exodus from Egypt. Immediately after the Israelites had crossed out of Egypt into the Sinai peninsula, the Hairy Ones were waiting and they attacked the refugees.

> So intense was their onslaught that, within days of the Exodus, the Israelites were forced to dig trenches, and cover them with branches so as to protect themselves from the stone-throwing barrage of the Hairy Ones.

Not only was Esau a Hairy One some six hundred years before, but he was not the only one of his kind among the Hebrews. There were many others like him, a sufficient number, in fact, to contest Jacob's subsequent leadership of the people. Ibn Ahron comments,

> But, just as they had threatened war against Jacob after their candidate had been defeated in the election of a Nomad chieftain... they had no intention of allowing Israelites to dwell in peace after the latter had made good their escape from Pharoah.

Strangely enough, although these Hairy Ones were sufficiently numerous to be a political force within the Hebrew tribe(s) in the time of Jacob in 1900 B.C., and numerous enough to be a serious menace immediately following the Exodus some six hundred years later in about 1330 B.C. (according to most opinions), *they did not harass the Israelites when the tribes were in Egypt.* Why not? And why were they waiting for the Israelites just outside the Egyptian border? Ibn Ahron confirms that the Hairy Ones were not a problem for the Israelites when the Hebrews were in Egypt.

> ...the Hairy Ones never harassed the Egyptian homeland.

Again, why not?

The simplest explanation is that these Hairy Ones were a fairly numerous, embarrassing and unwelcome part of Israel; a strongly Neanderthal-like group of relatives which was consistently denied representation and who were more or less ostracized, but who tried continually and unsuccessfully to assert their rights while following the more "normal looking" Israelites as pathetic and angry camp-followers. Now, could it be that the Egyptians refused admittance to this numerous "visible minority" within the Israelite host when the Hebrews pleaded for entrance into Egypt during a time of famine? Were the Hairy Ones simply left in the Sinai desert as the rest of the Israelites gained entry into Egypt? And, helpless and angry as spurned outcasts, did the Hairy Ones wait doggedly just outside the Egyptian border for the Israelites to come back eventually?

That seems to be the only explanation for the fact that the Hairy Ones plagued the Israelites both before and after their sojourn in Egypt but not while the Israelites were in Egypt. What about those Hairy Ones who must have been born to the mothers of Israel while the Hebrews were in Egypt? There must have been some of those Hairy Ones born because a fairly large number of these "throwbacks" had been produced continually from times long past. They were produced in significant quantity later. What happened to the ones born in Egypt? The only answer is that they must have been killed in infancy and very possibly by order of the Egyptian authorities. Was there some genetic reason why the *first male child* of an Israelite woman was more likely to be a "throwback?" Was this the reason, during the Israelite sojourn in Egypt, the Hebrews were subjected to infanticide of some *first male children*? It seems strange that the Hairy Ones ceased to be a significant part of Israel while the Hebrews were in Egypt.

If this notion has any medical or genetic validity, then we may ask: did the Israelites seek admittance into Egypt *not only* because of famine, but *also* to escape and cleanse themselves from the genetic curse of the Hairy Ones? Did they leave Egypt when the forced labor to which they were subjected became too onerous?

Whatever the validity of such speculation, the Hairy

Ones again became a genetic burden on the Israelites in the generations following Moses and the Exodus. As Ibn Ahron states, after the Exodus...

> As soon as things settled back to normal...the God of Israel decided to assert his power against the Hairy Ones. His real reason for doing this is that, in the days before He ever thought of speaking to Moses, they had rejected him as their diety, in favor of a certain Azazel. Thus, no sooner was the Holy Sanctuary completed, than Yehovah gave orders to Aaron, the High Priest, who, as we read in *Leviticus* 16:7 "took two of the Hairy Ones, and caused them to stand before the Lord. And Aaron put tags on the Hairy Ones, one tag for the Yehovah and one tag for Azazel. And the one which had been tagged for Azazel remained alive in the presence of Yehovah, who sent it off to Azazel (who lived in) the desert, that it might atone (for the sins of Israel, according to the commentators)." Lest the casual reader mistake the seirim of this passage with the "hairy goats" (serei 'izim) of the verses immediately preceding it—as did the famous but unhappily incompetent modern scholar S.L. Gordon—Leviticus Rabba, the most ancient exegetical works of Judaism, spells it out for him: "These seirim are none other than the Destroyers, the sons of Esau." The similar meaning of seirim and sheidim is among the ten best cases of authenticated synonymity in the Hebrew language.
>
> Thus, it may be understood that the "scapegoats" so glibly mentioned in English translation, and in countless works on anthropology and psychology, were, in fact, manlike creatures, with considerable biological, if not spiritual and intellectual affinity to the rest of mankind, and that Aaron, acting under orders from Above, committed an act which, to our eyes, looks like something very close to human sacrifice.

> The commentators are agreed in admitting this, but, as in the case of Jacob's deception of Isaac, they do not debate the morality of Aaron's obedience to the will of the Most High...Israel must turn out the Beast within, and sever itself from the sons of Esau, born of the same woman (Zohar).[25]

Rabbi Yonah Ibn Ahron's analysis here is sufficiently heavy with implications for Jewishness that no additional comment is needed. Perhaps it is necessary only to add that the Hairy Ones were sometimes called "Abominable Ones." The word for *abominate* in Hebrew contains the same root letters as the word *seirim* ("hairy"). It is remarkable that the Israelites of three thousand years ago had chosen the same word to describe Neanderthal characteristics as English newspapers and Eurasian Sherpas—"abominable." Is Western humanity's shame at its partial Neanderthal origins really so deep-rooted?

The supreme triumph of Israel is supposed to be "the long Jewish struggle toward monotheism," as absolute dominance of Jehovah in Judaism has been romantically called. Because "we" of the Western World have embraced either Judaism itself, or Christianity and Islam which largely derived from it, we naturally tend to consider monotheism as some sort of moral and spiritual victory over paganism and pantheism. However, we should stand back, objectively, and ask whether this is so, or whether monotheism is simply the result of a fairly strong racial proclivity? Does monotheism derive from actual spiritual superiority, or does it derive essentially from Neanderthal psychobiological maladaptations?

Any reader of the Old Testament knows that Jehovah cannot be described as gentle and easy-going. "He" demanded all of the godhead for himself—"thou shalt have no other gods before me." "He" successfully ousted women and the female principle from participation in divinity—from a biological perspective, Jehovah is the ultimate in divine sexual dimorphism, the female part no longer even exists! "He" was always a wrathful god, demanding absolute obedience and intolerant of anyone who did not tow his line. "He" meted out unbelievable punishments for anyone who did not follow his rules. Yet, there are numerous Jewish and Christian writers who point out the progressive and gentle nature of Jehovah and monotheism by smugly asserting that the God of Israel (as a rule) did not demand human sacrifices, while gods like Chemosh and Baal liked their victims. This is nonsense. Chemosh and Baal, Ishtar and Demeter were all satisfied with the occasional murders of sacrificial victims; Jehovah was satisfied with nothing less than genocide. "He" decreed the complete extermination of the Canaanites, Amalekites and even almost all of humanity itself (the Flood). "He" frequently threatened his own supposedly chosen people with near extermination. Compared with Jehovah, those "pagan" gods and goddesses were no match in terms of murder and mayhem.

In brief, Jehovah has behaved in the celestial realm just as Caucasoids have behaved on earth. *The Iceman Inheritance* presents a very defensible case for supposing that Caucasoid behaviors resulted from Neanderthal glacial adaptations. It seems reasonable that Jehovah is merely the psychological projection of the same glacial adaptations into the celestial, "metaphysical" and so-called spiritual realm. The *Bible* says that God made man in His image, but even 2,400 years ago Herodotus knew better. It is the other way around, man makes his gods in his image. Herodotus observed that the gods of the Ethiopians were "snub-nosed and black." Similarly, the God of Israel was made in the image of glacially traumatized psychology. Jehovah is the "spiritually living" image of a glacial God and if we must look upon him with fear, then we must fear ourselves (as others do).

It is not necessary here to discuss all of the myriad ways in which the Jews fashioned Jehovah in the image of Neanderthal psychology. The jealously, intolerance, violence, anti-feminism—all have been covered adequately, in general, in *The Iceman Inheritance* and outlined in part of this book. The situation must be clear enough to any objective reader.

Anti-feminism within Hebrew social and religious conceptions, transferred by a sort of cultural osmosis to, first, Christianity, and then Islam, is presently a topic of concern and controversy for major Christian churches. Many theologians and feminist writers have devoted themselves to freeing the "Church" from these early Judaic discriminations. The major Protestant churches are grappling with this issue, which is sometimes considered under the general heading of "Paulist doctrine." This is because Saul, a Jew who converted to Christianity (to become St. Paul), firmly cemented into fledgling Christianity some long-standing Judaic anti-feminisms which were in the nature of Jewish social norms or tradition, of the time, but which were not reflected in anything that Jesus ever said. These strictures are contained only within Paul's own prolific writings and injunctions.[26] The Paulist influence has plagued almost 2000 years of Christianity, and Protestant denominations are only now, and hesitantly, breaking free of it. In the mid-1970s, Rev. Allison Cheeke was ordained within the Episcopal Church in the U.S. but had a hard time finding a parish. Canada, typically a decade or more behind innovation elsewhere, is only beginning to confront the problem now. And it is being "solved" in a typically Canadian way of avoiding the real issue while "appearing to be progressive" and yet ensuring that no vested interest apple carts will likely be upended. The Anglican Church's policy is to ordain women and then send them to the parishes that no other priest wants: the high Arctic, Indian and Eskimo villages. When these women, under the pressure of sarcasm and ridicule by their male colleagues and superiors in the comfortable south, and experiencing privation, isolation and real physical danger, inevitably suffer emotional difficulties—then they are declared to be unfit for God's work and denied any chance of getting a parish in Toronto, Ottawa, Vancouver or any of the smaller towns and cities in the (more or less) habitable southern part of Canada.[27]

As for this issue within the Roman Church, change seems at least possible—but it will take a long time. And, possibly, change will come about through competitive pressure from the revival of Catharism in Europe, a heresy (from the Roman Church's perspective) that threatened the medieval Church like nothing since the barbarian invasions at the fall of Rome 1600 years ago. Neo-Catharism is a growing European reality and the reader is referred to *The Iceman Inheritance* (pages 137-140) which gives references for further research.[28]

As for the actual conceptual core of all this, Judaism itself, it is unlikely that women will ever be considered worthy of being rabbis or priests. If the imperative to assert and preserve male identity-in-time is as powerful among strongly Neanderthal-influenced populations as it seems, then only biological assimilation of the majority of Jews will ever permit female participation in the Godhead. Intermarriage of Jews with non-Jews to a significant degree is apparently the only possibility of disrupting the close and symbiotic relationship between the Judaic religion, "Jewish" culture and Neanderthalisms perpetuated by inter-Jewish marriages (genetic isolation). If Judaism becomes merely a religion, and both social norms and genetic selection are separated from it, then probably the conception of Jehovah and women's place in the interpretation of the divine would change. Something of this sort seems to be happening now, with a fringe group of Jews called the "Jews for Jesus" movement. Many or even most of these people are of Jewish descent, but there is a significant admixture of converted-to-Judaism gentiles (especially wives), and some change is evident in this small splinter group.

And what about Islam? In our time, at least, it seems to be, without doubt, the most anti-feminist of the three major Western faiths. From a purely biological perspective, this should not be the case. Moslems are all sorts of people: Black Africans, Orientals, some Neanderthal-Caucasoids in Europe (Albania). They should, long ago, have modified Islam to be less anti-feminist than, say, Judaism. And yet, as things stand now, it is more so. There is an explanation for this. Whereas people of all the above non-Western groups, plus true Arab "Semites," presently embrace Islam, it will be noted that the spiritual, traditional and military center of this religion is presently Iraq and Iran—a region which, ethnographically and geographically, is our old hotbed of lingering Neanderthal traits: the Caucasus-Elburz mountain complex. Ayatollah Khomeni came from the Zagros mountain village of Khoumein about 175 miles southwest of Tehran and about the same distance from the late Neanderthal site of Shanidar. The Ayatollah was fiercely Caucasoid, and perhaps a bit more: his nose and brow ridges betray strong Neanderthal traits. Perhaps his genetic blueprint, as much as or more than his spiritual orientation, dictated his Islamic fundamentalism complete with a large dose of anti-feminism and male-dominant patriarchalism . . . (Salman Rushdie, can we share accommodation?)

However, when it comes to Islam and "Arabs," we are in much the same muddle as with Jews. Islam is a religion. Most, but by no means all, Arabs as a genetic-cultural group adhere to this religion, so much so that the Moslem-Arab distinction has been muddled for them and us. It has also been muddled and made emotionally acute by military and geopolitical factors. The Arab culture-ethnic-linguistic world has been threatened by, first, the Euro-American exploitation (there's really no other word) of Arabic oil resources since the 1920s. Since 1948, that world has been further threatened and insulted by the artificial creation of Israel in the region and by continued external Western military support of Israel, both passive (supply of arms) and active (intervention by Western armies) which the Arab world cannot match. Further, Jewish influence over Western media, and over America's media in particular, has ensured that incessant propaganda in books, films and television (not to mention newspapers) has thoroughly brainwashed the U.S. public into a pro-Zionist/pro-Israel kind of zombiism.

The Arabs know full well that the American public, in

general, does not even grant that Arabs have a valid complaint and political position, much less begin to understand and appreciate the finer points of that complaint and position. This American ignorance has been created, nurtured and molded by writers like Leon Uris (*Exodus*, the novel) and Hollywood producers like Otto Preminger (*Exodus*, the film) and many other (more subtle, but also more incessant) pro-Jewish expositions on television, all disseminated with the assistance of Jewish finance.

The Arabs, like all other Western (Caucasian, Neanderthal-Caucasoid) people, never had a great deal of internal cohesion or mutual good-will. They fought and struggled among themselves, much like Europeans, in an historical situation that illustrated just how weak the supposed binding force of shared religion really was. Not very strong at all. Western military power, plus Jewish/Zionist propaganda to deploy that power, has changed that situation somewhat. Islam, the religion, is about the only thing that binds the beleaguered Arab world together, and it is not at all suprising that, under our military/propaganda onslaught in assertion/defense of Israel, that religion should be expressed in bottom-line, fundamentalist terms. When one is threatened with no foreseeable hope of redress, one becomes dependent for identity on very basic conceptions. That is what has happened in the "Arab World."

In fact, Islam tolerated more "pro-feminist" divergence than Judaism ever did when it was not threatened by our modern, Western steamroller of "oil exploitation-military force-Israeli propaganda." Medieval Mali, for example, was nominally Islamic in religion, but its women did not practice purdah and there were several other departures from the Ayatollah's ideas of right and wrong.[29] Westerners, and this includes the Jewish/Zionist/Israeli media-control complex, have created Islamic fundamentalism as a last defense of the Arab world.

However, these considerations are largely socio-economic, cultural and military in essence. They are not primarily biological. Being cultural and complex, the complexities can be argued back and forth eternally *in order to disguise the real psycho-biological issue that is relevant.*

And that issue is that the Hebrew creation of Jehovah, with His aggressive and anti-feminist personality, was unique in the Ancient Middle East and derived directly from a highly Neanderthaloid psycho-biological profile of His creators.

Let's talk about "idolatry."

An idol is, by definition, a visual image of something that is worshipped or adored. With the sole exception of the Hebrews, other peoples of the ancient Middle East were enthusiastic and talented creators of visual images of their gods and goddesses. The Hebrews, alone, regarded their God as an entity who could not, must not, be depicted visually. "He" was given a residence, the Ark of the Covenant, but it was an empty box. The invisible God resided therein.

This "abstract" conception of the Most High does not derive from any "higher conceptions" or "long struggle toward monotheism" of the Jewish people, but merely from the *known* (by archeological evidence) *lack of Neanderthal ability to create visual images.*

Neanderthals, and strongly Neanderthaloid populations derived mainly from them genetically, do not seem to have much visual artistic ability. That is not their strong suit.

What they do have, in terms of artisitic ability, is an appreciation of form, number and proportion. They excel in non-visual mathematical and numerical relationships. Maybe that is why we have so many Russian and Central European physicists, mathematicians and champion chess players—not to mention musical geniuses—but so few Central European painters of world renown?

It is tempting to suggest that the idea of visual arts (like painting) developed where there was form and color: the tropics and sub-tropics. It is also tempting to suggest that visual art took root and gained prestige along those corridors of post-glacial migration into Europe that carried people of long tropical tenure. Elsewhere, in "migration backwaters" we have the emergence of not very visually endowed artists, but people of great abstractional/numerical/proportional ability. Is it any wonder that Europe's greatest musical composers have come from the Russian-Central European area? Music is non-visual form and proportion, as is higher mathematics, physics and, at its best, engineering. Did this proclivity derive from the environmental fact that, during the glacial era, there was not much color—and that reality was black-and-white, ice-and-shadow, arranged in forms and numbers, like days and nights of the moon's phases?

Might this not plausibly explain one of Western man's intellectual-sexual quirks? Western man has not only alternately needed and rejected women (only to be expected from Neanderthal psycho-biology), but he has inexplicably also correlated women with the eternal and enduring, placed women on pedestals and worshipped them and also has violently rejected them. Now, could it be that this comes from the fact that human female periodicity coincides with the phases of the moon? A 28-day cycle, on average. Did this fact not give them (females) some "cosmic cachet" while, for other identity-giving reasons Neanderthal-Caucasoid men tended to denigrate and despise women? Being number-conscious, Neanderthal men must have been aware of the synchronicity of the female menstrual cycle and the moon's phases. Did this "coincidence" of the two cycles also add to the sexual-violence ambivalence suggested in previous pages? It must have and it is one of the reasons Western males tend to deify females (on the one hand), while almost simultaneously fearing/rejecting them on the other.

In brief, the ancient Hebrew prohibitions against representing their God visually was not so much a "victory of monotheism" as a simple, inherited, strongly Neanderthaloid inability to portray things visually. The abstract and intellectual nature of Jehovah was not so much a truimph of the human capacity for spirituality, as merely the lack of capacity of a certain group of humans to represent things visually. Similarly, I think that the see-saw ambivalence of these people regarding virtual sentimental worship of women may derive from the accidental correlation of the abstract lunar phase with the female physical menstrual cycle. Hence, the God of Israel, His psychology and demands, and the psychosexual maladjustment of strongly Neanderthaloid people in the process of evolving a complex culture or proto-civilization in the ancient Middle East.

The obvious differences between the Biblical Jews and their neighbors in the ancient Middle East derive mainly from their very strong Neanderthal physical and mental traits, not from any sort of moral superiority or special status as being God's chosen people. The Hebrews tended toward the creation of an abstract, invisible male god who would tolerate no competition as a reflection of Neanderthal psycho-biology. Other peoples of the ancient Middle East were pantheistic, represented their deities visually (idols) and were content to have the female principle represented more or less equally in the cosmic scheme of things: they had goddesses as well as gods. Ancient Egyptians, Sumerians and later Mediterranean peoples all developed sophisticated visual art based mainly on religious themes and inspiration. These works are the artistic treasures of our museums.

There cannot be much argument about all this. Biblical Jews originated in an area which, even today, produces populations with marked Neanderthal traits, an area which is rich in late Neanderthal archeological sites.

Although the Hebrew language is an adopted form of some Semitic dialect, the basic letter-names of the language betray a Caucasus origin. The proto-Hebrew tribes of the time of Noah and before spoke a version of primal Indo-European. Their own "before the Flood" traditions testify to genetic mixing between incoming (agricultural?) "men" and frightful beings from "on high" who came down to rape women. From these unions, the primal Hebrew heroes emerged.

Rabbinical tradition (the exegetical works), as well as the Esau story in the *Old Testament*, indicate a continuing, very strong Neanderthal genetic strain among the Israelite tribes—a strain that they tried to eradicate by various means.

There should be nothing odd or particularly shocking about this. If the Caucasus-Elburz region can still produce strongly Neanderthal people today, how much stronger must the Neanderthal presence have been in the distant past? Abraham and the Hebrew tribes arrived in Ur around 2000 B.C., 4000 years ago. Before that they took an unknown length of time to trek from "the mountains of Ararat" to Sumeria, and even before that they had lived in their mountains for another, unknown span of time. In the story of Cain and Abel and when we read about the "sons of those from on high," we are probably tapping into tribal memory that reaches back, at least, to 3000 B.C. and perhaps much earlier. At that time the Caucasus-Elburz Mountains must have had a much more marked Neanderthal character than now. It should not come as any surprise that the proto-Hebrews who emerged from this Neanderthal refuge 4000 years ago would have exhibited emphatic Neanderthal physical and mental traits. How could it have been otherwise?

But, the typical Neanderthal psychological trait is an inordinately high level of aggression, brought about by an overpowering urge to preserve identity in time. And it was this, and probably this more than anything else, that permitted the Jews to survive, against all odds, as coherent people in an area which was the natural borderland of huge and ancient empires. Palestine is the geographical bridge between Asia, Africa and Europe. Vast migrations and military campaigns have historically traversed this narrow geographical corridor. Most ancient tribes of this region (the *Bible* mentions Canaanites, Moabites, Amalekites and many others) were inundated, submerged and gradually assimilated into the vaster hordes from nearby continental land-masses. But not the Jews. They even managed, against all odds, to establish a suzerainty over the whole of Palestine, and to establish a kingship that was secure enough to win the awed recognition, for a brief two generations, of both Egypt and Mesopotamia. This was a remarkable feat, one that could only have been accomplished by formidable aggression and by an equal determination not to surrender identity. The Jews survived as a coherent people in the ancient Middle East, not because they were chosen and protected by God, but because they were strongly Neanderthaloid. Their conception of God, Jehovah, merely reflected the salient characteristics of this people.

V
The Khazar Conversion (Jewish Origins II)

From the time of Abraham (about 2000 B.C.) until some 1000 years thereafter, the Israelite tribes were led by clan patriarchs—inspired leaders who emerged in times of crisis (like Moses), prophets and judges. A little before the year 1000 B.C., however, the Jewish tribes in Palestine seem to have established themselves with sufficient security to conceive of a kingship and to want one. At this time, the recognized priest of the Lord was Samuel. Also, Samuel may have been a prophet—his status is somewhat unclear. He seems to have been divinely or mystically called to the priesthood (*Samuel* 3:1-21), but in any event, his position was recognized by the people. The Hebrews approached Samuel with their plea for a kingship. Samuel did not think it was a good idea. *Samuel* 8:4-5 . . .

> Then all the elders of Israel gathered themselves together, and came to Samuel unto Ramah. And said unto him Behold, thou art old, and thy sons walk not in thy ways: now make us a king to judge us like all nations.

Samuel listed all the bad things kings are prone to (*Samuel* 8:11-18), but still the people demanded their king. Samuel accordingly anointed Saul as the first Israelite king, but his only qualification for the job seems to have been that he was "head and shoulders" taller than any other Hebrew (*Samuel* 10:23). It must have been very difficult to establish a king over the formerly semi-independent and patriarch-led confederacy of Hebrew clans and septs, and Saul did not last. The *Bible* tells us that Saul displeased God by failing to massacre some Amalekites thoroughly enough. True, Saul had dutifully slain all the Amalekite men, women and children. He had burned or destroyed all of their meagre belongings. However, his men drew the line at killing off the Amalekite sheep and cattle, thinking that these herds could better be added to the Israelite wealth. Also, Agag, the chieftain of the Amalekites, had been taken hostage and not killed outright. Saul tried his best to convince Israelite soldiers to use the captured Amalekite sheep and cattle as burnt offerings to the Lord, and this was done in some instances but not in all—and we know Jehovah. *Samuel* 15:19-23 . . .

> Wherefore then didst thou not obey the voice of the Lord, but didst fly upon the spoil, and didst evil in the sight of the Lord?

> And Saul said unto Samuel, Yea, I have obeyed the voice of the Lord, and have gone the way which the Lord sent me and have brought Agag, king of Amelek, and have utterly destroyed the Amalekites.
>
> But the people took of the spoil, of the sheep and oxen, the chief of the things which should have been utterly destroyed, to sacrifice unto the Lord worthy God in Gilgal.
>
> And Samuel said, hath the Lord as great delight in burnt offerings and sacrifices, as in obeying the voice of the Lord? Behold, to obey is better than sacrifice, and to hearken (better) than the fat of rams.
>
> For rebellion is as sin of witchcraft, and stubborness is as iniquity and idolatry. Because thou has rejected the word of the Lord, he has also rejected thee from being king.

As for Agag, the helpless hostage, *Samuel* 15:32-33 says,

> Then said Samuel, Bring ye hither to me Agag the king of the Amalekites . . . And Samuel hewed Agag in pieces before the Lord in Gilgal.

The Lord had already located another royal candidate for Samuel to anoint: David, from the house of Jesse.[1] Aside from divine displeasure, it is difficult to figure out from the biblical account what really went on here. There seems always to have been some distinction between the "Israelites" (comprising 10 tribes) who at this time occupied the northern part of the Hebrew hegemony, and the two (larger?)

southern tribes who occupied what was known as Judah. Jerusalem, as a city (formerly Jebusi of the exterminated Canaanites),[2] seems to have occupied a geographical mid-point between Israel and Judah. Likewise, the tribe of Benjamin seems to have been "in the middle," too, with some of them being partisan to the north (Israel) and most being partisan to the south (Judah). Now, this first Israelite king, Saul, was a Benjamite and the city of Jerusalem was in land allotted to the Benjamites (*Joshua* 18:11-28). It is possible that Saul was initially chosen to be king as a "compromise candidate" between north (Israel) and south (Judah) that really satisfied no one. He seems also to have been a choleric man, given to fits of violence alternating with remorse—"mood swings" that would today be diagnosed as metabolic imbalance, possibly alcoholism.

In any event, Saul and David co-existed uneasily for many years, Saul alternately accepting David, at other times trying to murder him.[3] David conducted what can only be called a guerilla war against Saul between periods of truce. At last, David claimed and held the kingship of Judah, while Saul's heir ruled in Israel. A few years later, though, Saul's heir in Israel was murdered in his bed. David claimed the kingship of Israel as well. He made his capital, however, at the old mid-point of Jerusalem (a necessary political concession to bind north and south together?).

It is unclear how many "wives" David had. At least four are mentioned in the *Old Testament*. Two seem to have been acquired by murder of their husbands and immediate "appropriation" of the widows. Abigail came to David when the lord smote her husband, Nabal, ten days after David and Abigail had a little tryst[4]. The famous Bathsheba came to David after her husband, Uriah the Hittite (a mercenary soldier in David's pay), was sent into the heat of a battle with little hope of survival.[5] Both incidents have been romanticized in the *Bible*, but it is clear what happened. None of these women enjoyed any legal protection, or personal dignity, in any way commensurate with the traditional meaning of "wife." This is important to this analysis because it has significant ramifications in respect to Jewish-Islamic relations.

Although most North Americans are totally unaware of it, both Mohammed and Jesus claimed a plausible descent from Abraham. According to tradition, the Hebrews claim descent from Abraham and Sarah (including the family lineage of David which culminated in Jesus), while Mohammed descended from the union of Abraham and Hagar. Jews today rather smugly assert that while Sarah was the "wife" of Abraham, Hagar was only a *concubine*. So, *in Jewish eyes*, the line leading from Abraham and Hagar has an aura of illegitimacy about it. Ishmael, the son of Abraham and Hagar, was a "bastard," illegitimate, and Mohammed came from that background. The "Arabs," to use slightly incorrect nomenclature, come from a slightly "tainted" lineage and are not the "chosen people" like the Jews, who are "legitimate" descendants of Abraham and his wife.

This prejudice bears some investigation. Even in the time of David, about 1000 years *after* Abraham, we see that the women involved were simply *taken* (by fair means or foul) and no rites had to be performed to transform the woman into "wife." The *Bible* mentions no rites at all in the cases of Abigail and Bathsheba when David "took them to *wife*." One suspects that David simply had intercourse with them, willingly on their part or not, after murdering their respective husbands. *The American Heritage Dictionary of the English Language* (page 793) defines "wife" as a *woman married to a man*, with marriage inferred as being sanctified, publicly and in the eyes of divinity, by rites of mutual acceptance of some sort. Now, in the case of at least two of David's "wives," Abigail and Bathsheba, no rites of any sort are even hinted in the *Bible,* while the other two wives mentioned by name in *Samuel* were "given" as wives but without rites being specified.

Going back a thousand years to the time of Abraham, what on earth or in heaven could have differentiated between a *wife* (Sarah) and a *concubine* (Hagar)? And, what "rites" of marriage could have even been performed? Certainly not the Jewish rites because they had not yet been invented. Therefore, the distinction between Sarah and Hagar is artificial. This distinction was made, and written into the *Old Testament*, some 1500 years after the fact, during the Exile when the *Old Testament* was written down by scribes. Since the Hebrew kingdoms had just been destroyed, these scribes had a vested interest in recording a myth of Hebrew "chosen-ness." It was the only identity-preserving thing available to them at the time. So, Sarah was made into a "wife" and Hagar into a "concubine" so that their respective lineages would differ in cachet and "legitimacy."

But there is something further at stake here, something disguised and concealed for almost 4000 years. The son of Abraham and Hagar, Ishmael, was the first-born son of Abraham (*Genesis* 16) and should, by rights, have inherited Abraham's patrimony. But even more to the point, Sarah was the half-sister of Abraham (*Genesis* 20:12):

> And yet indeed she (Sarah) is my sister; she is the daughter of my father, but not the daughter of my mother; and she became my wife.

Now, we have to ask, why were Hagar and her son, Ishmael, cast out from Abraham's tent? *Genesis* 21:4-9 gives the answer: Ishmael was cast out because Ishmael mocked the late-born son of Abraham and Sarah, Isaac. Why would Ishmael mock Isaac? Perhaps we have an inkling of an answer if we skip ahead one generation to the birth of Esau. Isaac was Esau's father *and Isaac favored him.* Isaac liked the game-food that Esau brought back from his hunting expeditions.[6] Jacob had to win Esau's birthright by deceiving Isaac, with his mother's help, to pass himself off as Esau.

Could it be that Abraham was very strongly Neanderthaloid and that, in mating with his half-sister, they produced a highly Neanderthaloid offspring: Isaac, who later favoured highly Neanderthaloid Esau? Did Ishmael, the son of a true Semitic Egyptian (Hagar), mock the newborn Isaac *because of his primitive appearance?* And did Abraham cast Ishmael and Hagar out because he preferred Sarah's child who looked more like himself?

Then, much later when these traditions were written down and very heavily edited, and when the Neanderthaloid traits of the early Hebrews like Abraham had been much diluted, were the scribes 1600 years thereafter either genuinely confused about the motivations of Abraham, Isaac, etc., or, knowing the truth, did they feel the need to alter it? Until now, and due mostly to Rabbi Yonah N. Ibn Ahron's work, the real significance of Neanderthal admixture among the Hebrews has not been appreciated. It has never been a factor in scripture interpretations. But, indeed, it may be a very major factor in the later editing of what comes down to us as the *Old Testament.*

With the story of Abraham and his two mates, we face choices of varying appeal. First, we have to acknowledge that, circa 2000 B.C. among primitive nomads, there was no possible meaningful distinction between wife and concubine. Second, if there had been such marriage rites, by modern standards they could not have applied to Abraham and Sarah because they were brother and half-sister. By modern standards, their union was incest. Perhaps we have no right to apply modern standards. Egyptians also indulged in brother-sister marriages (with rites) *and did so in order to preserve genetic purity.* We accept this as their cultural norm without attaching modern morality to it—but if we extend the courtesy to Abraham and Sarah, and assume the same reason for their behavior, then they were married in order to preserve *their* genetic purity. Therefore, from Ishmael's mocking of newborn Isaac, and from Isaac's subsequent bias toward Esau, we are forced to accept the possibility that this purity was the preservation of highly Neanderthal traits. Third, we must accept that the true birthright should have gone to Ishmael, because he was the firstborn, and not to the latecomer, Isaac. And, fourth, by modern standards at least, Hagar was Abraham's only "decently possible" wife because (at least) incest was not involved.

Since Abraham and his subsequent lineage comprise the beginnings of "Jewishness," in which the Jews take pride by virtue of antiquity of this genealogy and its special favor in the eyes of God, we see that there are many unresolved mysteries right at the start of this long history—mysteries which, if resolved in the light of our present knowledge of genetic variation within the Hebrew tribes, might well reverse many long-standing myths, delusions and "smugnesses" of "legitimacy." Was Sarah successful in preserving "Neanderthal purity" for her late-born having the more modern Hagar-and-son cast out into the wilderness, while Rebecca was successful in undermining "Neanderthal purity" for her late-born (Jacob was the twin of Esau, but came second from the womb) because of Isaac's bad eyesight?

In any event, there is little justification for contemporary Jewish smugness about "legitimacy" compared to "Arabs" and Mohammed's lineage. The story of Abraham and Sarah on the one hand, and Abraham and Hagar on the other, is sordid and potentially embarrassing enough, so that any claim to Jewish "legitimacy" as the "chosen people" is absurd—*unless the claim rests solidly upon the idea of highly Neanderthaloid people being "chosen" by a highly Neanderthal concept of God* and of this claim being gradually evolved to apply to more modern men as Neanderthal genetic traits became diluted among the Hebrews.

Out of all David's "wives," the last (?), Bathsheba, was able to win for her son, Solomon, the heir to the Jewish kingship. According to Jewish tradition, somewhat sketchily confirmed by other Middle Eastern histories, Solomon was a good choice and managed to take Israel to her brief apogee of geopolitical power. It seems clear, at least, that Solomon achieved a polity uniquely more powerful than any previous Palestinian people under any previous king. Considering the imperial powers surrounding Palestine of the era, this was an achievement of no mean proportions. He held at bay Egypt in the south-southwest and "Assyria" in the east (Mesopotamia) in his lifetime. He constructed "the Temple" which, in its deepest recesses, was a repository for the Ark of the Covenant, the home of the Hebrews' invisible, abstract, omnipotent and intolerant god. The Temple *also* incorporated some never-forgotten tree lore from the ancient Indo-European memory of his people and he was (wisely?) willing to accommodate shrines of local Middle Eastern gods and goddesses within Jerusalem, if not within the "Holy of Holies" where the Ark reposed.

Naturally, later Jewish writers, while acknowledging Solomon's "temporal"[7] greatness, blame his god/goddess tolerance for the downfall of his Temple. This is only to be expected. But it is worth wondering if Solomon's polity could have existed at all without some concession to the beliefs of surrounding peoples, as well as some concessions to the "idolatry" practiced by a significant portion of the peoples whom the Hebrews had assimilated during their 1000 year-long conquest of Canaan. Solomon was supposed to have been very wise, and this suggests specifically one way he was wise (unappreciated by Jewish scholars). He was undoubtedly visited by the Queen of Sheba (Saba, in southeastern Arabia or Yemen?), and there is not much doubt that she was a beautiful Black woman. She repeatedly said so in love songs and poetry that, in Hebrew, can only be described as "rapping" (puns, allegories, metaphors and rhymed allusions). Given the performance of Solomon's father David, why should there be much doubt that this Queen of Sheba left Palestine pregnant? Why should there be much reason to doubt that this double-royal lineage did not eventually end up as the royalty of 20th century Ethiopia?

After Solomon, what might be called the unified

Jewish kingship dissolved back into its northern and southern component parts in some age-old and powerful distinction which no one can really grasp now. Since Solomon reputedly had "900 wives/concubines," we have to be careful about legitimacy here. But, on Solomon's death, his "son," Rehoboam, became king of the southern kingdom of Judah, while his "overseer," Jeroboam I, assumed the kingship of Israel.[8] Both of these rulers may have been equally legitimate.

With the Hebrews divided into two kingdoms, northern (Israel) and southern (Judah), they were not only often mutually antagonistic, but failed to offer concerted military opposition to various Egyptian, Assyro-Babylonian and Persian intruders from the Asian and African continental land-masses. After a thorough working-over by these Afroasiatic forces, the Hebrews were in no position to oppose the (first) Greek and (later) Roman intrusion from the European land-mass. What we now speak of as "Ancient Israel" actually lasted for *under 100 years* some 3,000 years ago! It was ruled by two kings: David and his son, Solomon. Furthermore, David came to the kingship sorely beset by Amalekites, Philistines, etc., and not in control of any Jewish homeland. It is on the basis of a very brief David-Solomon kingship controlling much, not all, of Palestine *for less than 100* years that one can properly speak of a "Jewish homeland"*—and it existed 3,000 years ago!*

Yet, on May 14, 1948, the modern state of Israel was proclaimed as a Jewish homeland with the backing of the United Nations under the UN supervision/intervention administered by Count Folke Bernadotte, and then later by Ralph Johnson Bunche, the first *Black* department head of the United States Department of State and (later, 1946)

Director of the United Nation's Trusteeship Division.

In cold, objective fact, one could better speak of Palestine as, say, a "Roman homeland." Romans conquered Palestine from the Maccabees, a Jewish sect, in roughly 71 B.C. and ruled it under a succession of puppet kings, of which the Herod dynasty of Christian fame was only one. Then in 70 A.D., after a Jewish revolt and the fall of the Jewish fortress of Masada, Romans had absolute control of Palestine until 640 A.D., when the Islamic Arabs over-ran the region. We have, then, a total of 711 years of Roman rule of this region; 141 years of rule over local puppet dynasties; and 570 years of direct rule. This compares with the king lists of Israel and Judah composed by the noted biblical scholar, Dr. D.W. Albright (see page 169).[9]

If we allow 100 years for the combined kingship of David and Solomon, and add that figure to each of the Jewish kingdoms, then we see that Israel (the northern Hebrew kingdom) endured for 301 years. Judah endured for 435 years. Both are far short of the Roman endurance of 711 years of puppet and direct rule over this same region. A "Roman homeland" claim to this region has better credentials than a Jewish one. Nor can it be said that the "puppet years" of Roman rule (total: 141) should be subtracted. In many cases, the kings of Israel and Judah were but puppet monarchs of some invading power: Egyptian, Mesopotamian or Greek. Nor can it be claimed that the Romans were somehow "more foreign" to Palestine than were the Hebrews. Both were foreign interlopers into this region. *Genesis* 12:6 tells us that when Abraham's tribes finally arrived in Canaan, "the Canaanite was in the land."

List of Kings of Israel and Judah			
Judah/King	YearsB.C.	Isreal/King	Years B.C.
Rehoboam	922-915	Jeroboam I	922-901
Abijah/"Abijam"	915-913	Nadab	901-900
Asa	913-873	Baasha	900-877
Jehosaphat	873-849	Elah	877-876
Jehoram (Joram)	849-842	Zimri	876
Ahaziah	842	Omri	876-869
Athalia	842-837	Ahah	869-850
Joash (Jeoash)	837-800	Ahaziah	850-849
Amaziah	800-783	Jehoram	849-832
Uzziah	783-742	Jehu	832-815
Jotham (regent & king)	750-735	Jehoahaz	815-801
Ahaz	735-715	Joash	801-786
Hezekiah	715-687	Jeroboam II	786-746
Manasseh	687-642	Zechariah	746-745
Amon	642-640	Shallum	745
Josiah	640-609	Menahem	745-738
Jehoahaz	609	Pekahiah	738-737
Jehoiakim	609-598	Pekah	737-732
Jehoiachin	598-597	Hoshea	732-724
Zedekiah	597-587	Fall of Samaria	722-721
Fall of Jerusalem	587		

From the above king-list, we see that the northern kingdom (Israel) succumbed first to foreign invaders. The capital city of Israel was Samaria. This was captured by Assyria in 722-721 B.C. and the population was carried off into slavery.

Judah survived a little longer, but was over-run by Nebuchadnezzar, a king of Babylon, in two campaigns (586 and 597 B.C.). Thousands of captives were carried off to Babylon—the so-called "Babylonian Captivity." This largely destroyed Jewish national cohesion because of the loss of both Israel/Samaria in the north and of Judah/ Jerusalem in the south. This is the so-called "Exile," but all records of it were preserved by the Jews taken to Babylon by Nebuchadnezzar. The Jewish kingship was ended. In its place, the "teachers" of Jewish history and tradition, or rabbis, took over the leadership of Jewish social custom, history and religious interpretation. This, in essence, is the "Judaism" that exists today—that which was preserved for a displaced and conquered people—their history and tradition.

However, in 539 B.C., an Indo-European conqueror named Cyrus ("The Great"), founder of the Persian Empire, conquered Babylon . Around 538 B.C., he freed the Jews taken to Babylon some 59 years previously by Nebuchadnezzar. These exiles were allowed to return to Jerusalem, where they re-built the Temple—but not, of course, with the grandeur that King Solomon had achieved. With these refugees, or "returning exiles," went Ezra, a scribe, and it is to him that the *Old Testament* in its present form is ascribed. He, supposedly, codified (wrote down) the *Old Testament* as well as many of the Jewish traditions that form the so-called "exegetical works." We tend to think of the *Old Testament* as ancient writings and of the *New Testament* as something much more recent, but, from a purely archeological point of view, the age difference is not so very great. *As an artifact*, the oldest example of the *Old Testament* comes from the so-called Dead Sea Scrolls which date from about 70 B.C.[10] The oldest example of the *New Testament*, in the form of the Gospel according to Thomas, comes from the Nag Hammadi texts of about 70 A.D.[11] The difference is 150 years, no more.

Nonetheless, from such recent "ancient" texts, the so-called Massoretic scholars have codified the Pentateuch and the Torah, the essence of what non-Jews call the *Old Testament.* Most of the original versions of the exegetical works are likely to have been written down in Ezra's time (circa 450 B.C.) and so can safely be assigned a truly old traditional origin, even if they had been subjected to rather biased editing during the Exile and afterward. But they are recent in actual written form.

Meanwhile, Alexander of Macedon ("The Great") effectively gained control of Palestine after his victory at Gaugamela (331 B.C.). This imposed a sort of pan-Eurasian "Hellenism" upon the Jews for 191 years until the Maccabees revolted in 141 B.C. In brief, and in some over-simplification, from the time of the re-building of Solomon's Temple around 530 B.C. until the final destruction of the Temple by Romans in 70 A.D. and the fall of the Jewish fortress of Masada in that same year, the evolution of Judaism was in the hands and minds of scholars, historians and interpreters of Jewish history (ie. "rabbis"), *and it has been ever since.* They distrusted kings, perhaps with good reason, as Samuel had distrusted them circa 1000 B.C. and also distrusted the tribal tradition/yearning for a king/messiah who would restore the old David/Solomon greatness. This very understandable distrust of kings is what Jesus of Nazareth encountered.

So, as of about 70 A.D., we have the following situation. The ten "lost tribes" of Israel (northern kingdom) had been carried off into captivity by the Assyrians as of 722-721 B.C. (the fall of their capital, Samaria) and have never been heard from since. But the restored people of Judah survived in Palestine until about 70 A.D. when many of them left because of complete Roman subjugation of the region. Jews dispersed all over the Roman world, but it is clear that a large number of them settled in southern France near the modern city of Marseilles and throughout the large Rhone River estuary. Although there was an influx of Jews into this Rhone River region after 70 A.D. and the fall of Masada, Jews had been established there for a very long time. No one can say exactly how long, but perhaps there was a significant number of Jews in southern France even as early as the time of David and Solomon around 1000 B.C. Traders from the Palestinian cities of Tyre and Sidon had sailed into the Western Mediterranean by that date and had begun a lively bronze trade based largely on Spanish copper and (some say) tin from Cornwall in Britain.[12] Colonists from Phoenicia founded Carthage in 814 B.C., such was the wealth of trade in the Western Mediterranean. There can be no doubt that many Jewish merchants and traders participated in this Phoenician-Carthagenian commercial exploitation of the Western Mediterranean. In the first place, some of the Hebrews seem to have been closely related to the Phoenicians on the coast of Palestine. The Benjamite tribe, for example, refused to join in a war against the "sons of Baal" (the Phoenicians of the Palestinian coast) and fought a civil war against the rest of the Israelite tribal confederacy in order to assert their position in this matter (*Judges* 19 and 21). We must remember that the Israelites were, to some degree, a composite people. Not all the tribes originated from Abraham's migration but seem to have been assimilated later by Hebrew conquest. The Tribe of Benjamin supposedly derived from the "youngest son" of Jacob and Rachel (*Genesis* 35:18-24), but this may be just a poetic or allegorical way of saying that the Benjamites were the last major ethnic addition to the Hebrew confederacy of tribes. They may have been Phoenicians, or at least closely related to the coastal Phoenicians, because they refused to go to war with the other Israelites against them. When the Benjamites lost this civil war, their survivors were carried beyond the reach of Israelite retribution to Greece[13] and presumably on Phoenician ships.

In any event, at least some Jews were closely related, by culture and probably kin, to the Tyrians and Sidonians of the coast. They undoubtedly had some involvement with the Phoenician-Carthagenian commercial activities in the far west. Even back in 1000 B.C., the mouth of the Rhone River was a port. Some Jews were probably living there even at that early date. Around 725 B.C., the Greeks penetrated into the Western Mediterranean, planting colonies, of which the first (in that year) was Kyme, later called Cumae, the mother-city of the modern Naples in Italy. Roughly a century later, circa 600 B.C., the Greeks founded Massilia, modern Marseilles.[14]

At the very time when Assyria was threatening Samaria of Israel (circa 722 B.C.) and at the very time when Nebuchadnezzar was threatening Jerusalem of Judah (circa 597-587 B.C.), Greeks were establishing a rather sophisticated civilization in the Western Mediterranean. It cannot be reasonably doubted that far-sighted Jews in Palestine, seeing the way things were going for Israel and Judah, emigrated in time to avoid the onslaughts of Assyria and Babylonia, respectively. A large number must have resettled in the new Greek colonies in modern France and Italy or, for that matter, in Carthagenian cities like Tartessus in Spain. And, in 530 B.C., Carthage annexed all of Spain, opening up many towns, cities and villages to commerce and immigration. There can be little doubt that Jews were a part of this spread of Palestinian (Phoenician-Carthagenian) culture into Iberia. Southern France and Spain were well-known to the Jews. In France, around the Rhone estuary, there was a town called Sion. The region's ancient Salic Law has been traced back to Judaic law, a count of the region was named Miron Le Levite and a brother of King Clothair II was named Samson.[15] Spain was called *Sepharad* in Hebrew, and even before the troubles of 70 A.D., many Jews lived there. In fact, because of the large Jewish community in Spain and southern France from early times, that region became a natural refuge for the large number of Jews fleeing from Roman massacres of Jewish rebels.

The word *Diaspora* ("The Dispersion"), applied to Jews, has various and puzzling definitions among numerous writers. Some, like Howard Fast, seem to apply it primarily to the great flood of refugees following the fall of Masada to Roman forces in 70 A.D. Others seem to apply it to the series of conquests that began in 722-721 B.C. with the fall of Samaria and the northern kingdom of Israel, continued in 587 B.C. with the fall of Jerusalem and the southern kingdom of Judah, including the Babylonian Captivity of roughly 587-538 B.C. (about 60 years) and ending with the various Graeco-Roman incursions that culminated with the fall of the Masada fortress in A.D. 70. All of these catatrophic events dispersed Jewish survivors, particularly the disaster of 70 A.D. because the Romans expelled Jews from Judea (formerly Judah) and destroyed the Temple.

One very important point must be made about this dispersion because it has some bearing on the interpretation of later events. When Assyria captured Samaria and the northern kingdom of Israel about 722-721 B.C., the northern ten tribes of Jews were carried off to an unknown, but terrible, fate. They have never been heard from since. No hint of their tragedy or any subsequent history has ever surfaced; they must have been totally exterminated. It may seem that these ten lost tribes must have represented the majority of Hebrews in Palestine, but this was not the case. The southern kingdom of Judah, consisting chiefly of two tribes (Judah, Simeonites, plus some portion of Benjamites),[16] was always the stronger kingdom and the core of Jewishness. So, the loss of the northern kingdom, though tragic, actually left the majority of Jews and the essence of Hebrew culture intact. When the southern kingdom of Judah, in turn, was conquered by Nebuchadnezzar about 140 years later, the Hebrews were carried off as captives to Babylon, but they were not exterminated. About 50 years later (538 B.C.), Cyrus of Persia captured Babylon and allowed the Jews (including Ezra, the scribe) to return to Judah and Jerusalem to rebuild their Temple. But it seems that not all of the Jews in Babylon elected to return to Judah; some had put down roots and remained.

The important point is that the "ten lost tribes of Israel," taken by Assyria, were truly lost, probably exterminated. No descendants of them have ever surfaced, no history of them has ever come to light. Not even any folklore or legends about them, or what became of them, survive in even garbled and distorted form. As for the Jews who elected to remain in Babylon when Cyrus released them from their captivity, there cannot have been very many of them. If they later radiated outwards from Babylon through peaceful trade and migration, their natural routes would have been southward and along the Persian Gulf to Arabia following well-established and very ancient trade and culture centers. In fact, there are small pockets of Jewish populations in Yemen, India and Ethiopia today[17] (a point that is crucial to our discussion).

When Rome expelled the Jews from Judea in 70 A.D., they dispersed according to the opportunities, resources and limitations of individuals and families. They were sprinkled throughout the Ancient Mediterranean world. While a significant concentration of them ended up in Alexandria and Egypt, the majority seem to have gone to southern France and Spain where there was already a large and thriving Jewish presence. They seem to have survived the fall of Rome (410 A.D. with the conquest of Alaric) and to have forged close links with the successors of Rome in Gaul (ie. France), the Sicambrian Franks. These Franks established the Merovingian kingships in post-Roman France. Jews were so much a part of these Frankish dynasties during the 5th-7th centuries that one modern historian has stated that the roots of French nobility are Jewish![18] Since the Salic Law of the Merovingians and southern France does trace back to Judaic Law, and most scholars are agreed on this, Jews must have been a very influential part of the local culture.

The religion of Islam now comes into the picture in a big way. Islam counts its beginning from Mohammed's flight from Mecca to Yathrib (now Medina, Arabia) in the year 622 A.D. This is called the *Hegira*, and all Islamic dates are counted from this event. Islam grew at phenomenal speed. Its supporters and adherents launched a *jihad*, or "holy war," to spread their faith all over the world. Jews participated in this spread of Islam and were carried along on the Moslem tide. In general, the early Moslems valued

Jewish literacy, culture and administrative know-how. Jews accompanied Moslem armies, not as despised captives, but as honored and trusted allies. Many Jews were in the entourage of Tarik, the North African berber chieftain who began the Moslem invasion of Spain. Tarik defeated King Roderick of the Visigoths in A.D. 711 near Gibraltar (*Jabal Tarik*—"Tarik's Mountain"). Thus, in that year, Spain had yet another infusion of Jews, adding to the many who had previously settled in Spain and southern France both before 70 A.D. and afterwards. Although the Moslem advance was stopped in France by the Battle of Tours (A.D. 732), 140 miles southwest of Paris, for many years southern France—and almost all of Spain—remained under virtual Moslem ("Moorish") control. Indeed, a rather unique culture, or true civilization, developed in the Pyrenees Mountains that straddle northern Spain and southern France. This is usually called Provencal Culture, far in advance of the rest of Europe in arts, letters and commerce, and about equal parts Moorish, Jewish and Celtic. This is where the Cathar heresy evolved.

The Jews in this region of Spain and France—resulting from various immigrations—became known as Sephardic Jews from the Hebrew word for Spain, *Sepharad*. Sephardic Jews can trace a genuine ancestry back to Palestine. They, along with colleagues in Alexandria, undertook the preservation of Jewish scripture, traditions, law and exegetical works by re-casting them in an evolving type of Hebrew language, and in other languages. Jewish scholarship gained respect and indeed became almost legendary in both the Islamic and Christian worlds during the Middle Ages. Jews played a very considerable role in the development of the Renaissance, but here we will mention only their contribution to the translation of Greek works into local European languages and Latin (sometimes via intervening Arabic versions). Theirs was a crucial contribution to the rebirth of European learning.

Very gradually, Jews in Spain and southern France began to penetrate into the rest of Europe.[19] This was a very slow process. To begin with, there were few Sephardic Jews who had any inclination whatever to penetrate very far into what was an extremely barbaric and backward medieval Europe. When some did gradually move into northern France and modern Germany, they often succumbed to Christian mobs and the Black Death. For Jews moving into medieval Europe, the Black Death held special terror; not only did they die from it (like anyone else), but they were frequently blamed for it. Thus, the Jews that managed to escape the plague very often were lynched by hysterical townspeople.[20] Modern Jewish historians have concluded that there is no evidence of Jews in Germany much before 1225 A.D. in any significant numbers.[21] The Black Death of 1347-8 decimated many of these. Nonetheless, by about 1360 there were some Sephardic Jews in eastern Germany, including even some rabbis.

At that time in Poland, Casimir III ("The Great," 1310-1370) was in the process of trying to transform Poland into a more modern example of European feudalism. He looked toward the west for his cultural inspiration, and found it in Germany—which was considerably more advanced than Poland. Under the so-called Magdeburg Law, Casimir invited Germans to resettle in Poland in large numbers. These immigrants built up-to-date medieval cities, such as Krakau (Cracow) and Lemburg (Lvov),[22] and were granted a great degree of self-rule. It is said of Casimir that he "found a country of wood and left a country of stone."[23] Casimir founded the first university in Poland, in Cracow in 1364, and German was for a long time the official university language.

With these German immigrants into Poland came a certain number of German-speaking Jews, and rabbis. In the Polish semi-wilderness, they encountered something that astonished them deeply. They encountered—Jews! Hundreds of thousands of Jews. More Jews than in all the rest of Europe combined. More Jews than had been in Palestine at the time of Jesus.[24]

And they were strange Jews. They spoke a strange language called Yiddish, of which some words were obviously a very primitive form of some Germanic tongue (but nothing like any sort of German spoken to the west)[25]. Some words were basic Hebrew, others were like Russian or Ukrainian, while some words were incomprehensible to the incoming European Jews. These strange Jews wore a skull-cap they called a *yarmolka*, and their rabbis wore fur-trimmed hats and a kaftan straight off the Russian steppes. They had strange eating habits and were fond of preparing white-fleshed fish a certain way. Theirs was a quite primitive form of Judaism and the most primitive among them, called Karaites, rejected even the idea of rabbis. They wore side-locks of hair and called their towns shtetls.

Who were these Jews? And where on earth had they come from?

The origin of Eastern European Jewry, or "Ashkenazim,"[26] was debated for along time. It was something of a mystery, even for Jewish historians, because these Ashkenazim could not be attributed to any known Jewish migration of the Diaspora. They were not remnants of the "ten lost tribes" because their Old Testament read just like the Sephardic one, the one supposedly written down by Ezra, and Ashkenazim traditions also lamented the loss of the ten northern tribes from the perspective of the scribes of Judah. They had no myth or folklore linking themselves with the lost tribes.

Could they have descended from those Jews who elected to remain in Babylon after Cyrus released them from their captivity? No. First of all, there were simply too many Ashkenazim, millions of them even in the 14th century, and not nearly enough Jews had remained in Babylon to account for them—even if these Babylonian Jews had somehow migrated to Eastern Europe *en masse* at some point in the dim past. But it was well known that there had not been any mass migration of the Babylonian Jews. Most had stayed in Babylon. Their descendants formed the large Jewish community that had existed in Bagdad (75 miles to the north) during Islamic times, in fact, up until modern Arab-Israeli conflicts. There had naturally been some radiation from Babylon in all the intervening centuries, but most of these Jews had followed trade, commerce and culture southeast toward Arabia and Yemen, India.

Certain non-Jewish historians from Central and Eastern Europe had always maintained that these numerous Jews had come from Russia[27] from the region north of the Black Sea. Since they were very numerous in the Poland of Casimir III's time (circa 1360 A.D.), and also equally numerous in what is now Latvia, Lithuania, Estonia and the Ukraine, they must have entered Eastern Europe earlier.

Thanks primarily to the work of Jewish historian, A.N. Poliak of Israel, the mystery of the Ashkenazim has been largely cleared up with a plethora of historical documentation. Poliak's first major work was an article in *Zion* (1941), in Hebrew, published in Jerusalem under the title "The Khazar Conversion to Judaism." His second work was his book *Khazaria—The History of a Jewish Kingdom in Europe* (in Hebrew, Mossad Beliak, Tel Aviv, 1951). Poliak's purely historical and bibliographic research has been fully supported by the excavations of the Soviet archeologist, M.I. Artomonov, who located the ancient Khazar capital and recovered Khazar-Jewish coinage and other artifacts. Artomonov published *Khazar History* (in Russian, Leningrad, 1962).

Needless to say, no one has seemed particularly anxious to make these works available in English to North Americans. However, the noted Jewish writer, Arthur Koestler, did write a popular account of this research (and other research relevant to the issue) called *The Thirteenth Tribe* (Hutchinson & Co., London, 1976). It is a very difficult book to find in libraries in North America. Copies tend to "disappear" from library shelves and are not replaced.[28] The facts have been suspected for at least a century and a half. Poliak presented the historical documentation as early as 1939 and Artamonov proved the case with actual excavation after World War II (paper presented in 1946).

It all boils down to this: the Askenazim have no genetic or historical origin in Palestine at all. They are descendants of Russian steppe tribesmen who were converted to Judaism about 740 A.D. because of political convenience. These steppe tribesmen were known as Khazars—the word *cossack* probably derives from their name—and they welded together a formidable empire that lasted for at least two centuries. However, around 965 A.D., the Khazar army was defeated by Sviatoslav, Duke of Kiev, and Khazar power declined. Their domain was much reduced, but what really finished the Khazars as a steppe empire was the invasion of Russia by Genghis Khan in 1218 A.D. (and subsequent Mongol campaigns). By 1250 A.D. they, along with a disparate flood of non-Khazar and non-Jewish refugees, had been pushed by the Mongols into Eastern Europe. They lost their cohesion as Khazars, but retained their Judaism, their language and many of their steppe traditions. They settled themselves into towns much like they had on the steppes of Khazaria, which they called *shtetls* and, by the time of Casimir III of Poland, they had settled down as "Ashkenazim"—Eastern European Jews who so astonished the incoming Jews and rabbis of true Sephardic origin.

Why did the Khazars convert to Judaism in 740 A.D.?

In a way, it was again the rise of Islam that catapulted the steppe tribe of Khazars from virtually utter obscurity into some prominence 1300 years ago, and into much controversy today. Within a century after the Hegira, Moslems had conquered Palestine (circa 640 A.D.) and had penetrated even further in Eurasia, toward the Caucasus, so that by 641 or 642 A.D. Arab Islamic forces conquered Persia. This new outpost of militant Islam obviously threatened and encroached upon the Christian Byzantine Empire—what had formerly been the eastern half of the old Roman Empire. Now that Rome itself had been sacked repeatedly by barbarians and could no longer be considered the capital of any empire, Byzantium had become, in practice, completely independent. Ties with Rome, such as they were, were not only weak but not particularly cordial. By 640 A.D., there was a definite and irrevocable difference in language between Rome and Byzantium, with the

Byzantines having long since reverted to their native Greek. There were also important differences in culture and custom, and departures from Roman orthodoxy in matters of religion. In brief, with the collapse of Rome in the west, Byzantium had evolved into an independent empire. *Byzantine* as a word has come to mean "corrupt," "decadent" and "rotten with intrigue," and the politics of this city (Constantinople)[29] and empire were often just that. On the other hand, this empire survived with cultural cohesion until 1453 A.D. Although generally viewed (by Romans, and later Europeans) as effete Greeks without the Western knack for warfare, Sir Charles W.C. Oman, in his monumental *A History of the Art of War in the Middle Ages* (Metheun, 2 vols., London, 1924), gives the Byzantines high marks for military research, tactics, strategy and administration. It was this that ensured Byzantium's lengthy survival. As early as about 400 A.D., Byzantium had developed an answer to the Gothic-Slavonic horse warriors who threatened to overrun the empire in successive waves, all coming from the depths of the Russian steppes. This answer was the heavily armored mounted archer. They were called *cataphracti* ("armored ones") and were deployed in troops of 100 horsemen. It was said, and proved often enough, that any Greek general with a thousand *cataphracti* at his back could defeat any horde of steppe horsemen. So, for about 250 years, from about 400 A.D. until about 640 A.D., Byzantium basked in some security; it kept the wild steppe tribes of Russia (Khazars, Bulgars, Magyars, Alans and others) thoroughly cowed. Byzantium did not have to worry very much about incursions by the wild Caucasus tribes (Georgians, Kurds, Armenians and others) because, although these people were fierce enough, they fought among themselves and kept themselves, and each other, confined to the mountains on the far eastern border of the Byzantine Empire.

Byzantium had settled down to enjoy delicious intrigue, a refinement of cultural decadence seldom matched in world history, and great wealth derived from being Europe's door to Far Eastern trade (via caravans from across Asia). This happy situation changed in 640 A.D. with the Islamic conquest of Persia. Suddenly, there was major, aggressive, and equally cohesive, military power right next door to the Byzantine Empire. The balance of power had changed. Drastically. In this new geopolitical ball game, the steppe tribes suddenly became more important. Particularly, the Khazars became important because they were the most powerful steppe tribe in the region between the Black Sea and the Caspian Sea. Not only were they an important link in the long caravan route to the Far East, but they now held the balance of military power. Not powerful enough to challenge either Byzantium or the new Islamic Persia on their own, the Khazars, allied with either, could ensure the defeat of the other. That being so, Greek Orthodox Christians from Byzantium and Islamic missionaries from Persia immediately launched energetic attempts to convert these Khazars. Religion was the obvious way of ensuring military and political control because the Khazars, if successfully converted to either Christianity or Islam, would fall under the spiritual control of Byzantium or Persia.

It is not clear at all what sort of religion the Khazars originally possessed. It was probably some form of shamanism, but it is very clear that the Khazar leaders fully appreciated the new situation. They knew well enough that their tribespeople might succumb to the more cultured and refined religious beliefs of the two sets of competing missionaries. At the same time, the chieftains (at least) realized with utmost clarity and *realpolitik* that they could not adopt either Christianity or Islam without losing much of their Asian trade profits and all of their military independence.

But some Khazar chieftain discovered Judaism and was intelligent enough to perceive that it was, in some ways, the parent religion of both Christianity and Islam. Judaism presented a solution to the Khazars' geopolitical problems. British historian, J.B. Bury, put the entire situation succinctly in his *A History of the Eastern Roman Empire* (London, 1912, page 401):

> The religion of the Hebrews had exercised a profound influence on the creed of Islam, and it had been a basis for Christianity; it had won scattered proselytes, but the conversion of the Khazars to the undiluted religion of Jehovah is unique in history . . .

And elsewhere, Bury (Ibid., page 406):

> There can be no question that the ruler was actuated by political motives in adopting Judaism. To embrace Mohammadanism would have made him the spiritual dependent of the Caliphs, who attempted to press their faith on the Khazars, and in Christianity lay the danger of his becoming the ecclesiastical vassal of the Roman Empire. Judaism was a reputable religion with sacred books respected by both Christian and Mohammadan; it elevated him above heathen barbarians, and secured him against interference of Caliph or Emperor. But he did not adopt, along with circumcision, the intolerance of the Jewish cult. He allowed the mass of his people to abide in their heathendom and to worship their idols.

Bury's comment about the "intolerance of the Jewish cult" suggests more than a hint of "anti-Semitism." The fact of the matter is that the Khazars first came into contact with Jews because of enthusiastic anti-Jewish persecution being practiced in Byzantium. Intolerance of Jews came to be a public issue under Justinian I (A.D. 527-565), became vicious under Heraclius in the 600's, continued under Leo III in the 700's when the Khazars were converted and was maintained in the 9th century by Basil and Leo IV (and by Romanus in the 10th).[30] Leo III, who ruled Byzantium during the twenty years just preceding the Khazar conversion, "attempted to end the anomaly (of the tolerated status of Jews) by ordering all his Jewish subjects to be baptized."[31] And, how was this coersion accomplished?

> How did they force them? Anyone refusing to accept their erroneous belief was placed in an olive mill under a wooden press, and squeezed in the way the olives are squeezed in the mill.[32]

Those Jews in Byzantium who wished to retain their faith, yet wished also to avoid the Olive Press of the Lord, fled. They could not flee southward, further into the heart of Byzantium. They could not go east because of the rather fanatical profile of frontier Islam just established in Iran. Who would go west into the wilds of the uncivilized Balkans? The only possible haven lay to the north. Khazar country. And there they found a pagan people, who may have been both fierce and primitive, but who exhibited a good deal more religious tolerance than those who worshipped either the gentle Christ or Allah the Compassionate.

So, because of Byzantine persecution, a few Jews came among the Khazars and the chieftains, at least, learned of Judaism—enough to bring off the politico-religious coup of history.

The actual Khazar conversion to Judaism occurred about the year 740 A.D. A detailed but obviously biased account of the event is offered in al-Bakri's *Book of Kingdoms and Roads* (about 1050 A.D.):

> The reason for the conversion to Judaism of the King of the Khazars, who had previously been a pagan, is as follows. He had adopted Christianity (another Arabic writer, Al-Masudi, says it was Islam that had been briefly adopted by the Khazar king—Author). Then he recognized its falsehood and discussed this matter, which greatly worried him, with one of his high officials. The latter said to him: O king, those in possession of sacred scriptures fall into three groups. Summon them and ask them to state their case, then follow the one who is in possession of the truth.
>
> So he sent to the Christians for a Bishop. Now, there was with the king a Jew, skilled in argument, who engaged him in disputation. He asked the bishop: "What do you say of Moses, the son of Amran, and the Torah which was revealed to him?" The Bishop replied: "Moses is a prophet and the Torah speaks the truth." Then the Jew said to the king: "He has already admitted the truth of my creed. Ask him now what *he* believes in. So the king asked him, and he replied: "I say that Jesus the Messiah is the son of Mary, he is the Word, and he has revealed the mysteries in the name of God." Then the Jew said to the King of the Khazars: "He preaches a doctrine which I know not, while he accepts my propositions".... Then the king asked for a Muslim, and they sent him a scholarly, clever man who

> was good at arguments. But the Jew hired someone who poisoned him on the journey, and he died. And the Jew succeeded in winning the King for his faith, so that he embraced Judaism.[33]

Of course, the Jew did not have to bother poisoning anyone. The Islamic mullah would have fared no better than the Christian bishop. Both Christian and Moslem accept the truth of the *Old Testament*, while the Jew "knows not" the doctrines of the *New Testament* or the *Koran*. The king, politically motivated in any case, saw the wisdom of accepting the "common denominator" (as Arthur Koestler put it).[34]

Thus, were the Khazars converted to Judaism sometime around 740 A.D. Now, what precisely does this mean? It means, most probably, that the king and the notable tribal chieftains following his lead immediately accepted Judaism, but that the mass of the people did not. Nonetheless, because of prestige of the new religion among the nobles, and because of expectations coming from the top down, it seems probable that ordinary Khazars adopted Judaism at a gradual, but steady, rate. Within a bit more than two centuries, say, at the height of Khazar power around 950 A.D., it seems reasonable to suppose that most Khazars had become Jews. Certainly, by the 10th century, an Arab chronicler, Mugaddasi, could observe "in Khazaria, sheep, honey and Jews exist in large quantities."[35]

This personal comment is verified by archeological finds. Upon the Khazar adoption of Judaism, the Hebrew language and script were adopted by the Khazar administration in the capital city, Khazaran-Itil, on the Volga.[36] Coinage was minted there, and the inscriptions are in the Hebrew language and script. The Hebrew language that was actually spoken, however, was not only a rather primitive form of the language but was probably only spoken as a customary tongue by the Khazar upper classes. Most Khazars probably continued to speak their own native tongue, but one increasingly influenced by what might be called "pidgin Hebrew." The Hebrew script, however, seems to have been adopted widely in order to write non-Hebrew languages—much like the Roman alphabet was adopted in Europe to write a variety of local tongues, some of which were not very closely related to Latin at all (English, for example). In Eastern Europe, archeologists have found the same thing: non-Hebrew, non-Semitic *languages* transliterated into the prevailing Hebrew *script.* Two tombstones found in Crimea are inscribed in Hebrew letters, but the language is unknown and they have not been translated.[37] Influence of the Hebrew script was so powerful that, even as late as the 12th and 13th centuries, many Polish silver coins are found using the Hebrew alphabet to transliterate a Polish-language dedication: *Leszek krol Polski*, "Leszek King of Poland."[38]

So, on the basis of archeological finds, Mugaddasi's observation is fully supported: there were a lot of Jews in Khazaria—at least, there were a lot of people who professed a basic form of Judaism appropriate to their own rather primitive "proto-civilization." They spoke, to varying degrees, a simple form of Hebrew liberally laced with their original language, but almost universally adopted the Hebrew script (because they had no other) in order to express their speech in written form.

How many Khazars? How many Jews? No one can say with any certainty. Arab accounts of Khazar-Moslem

Figure 4:

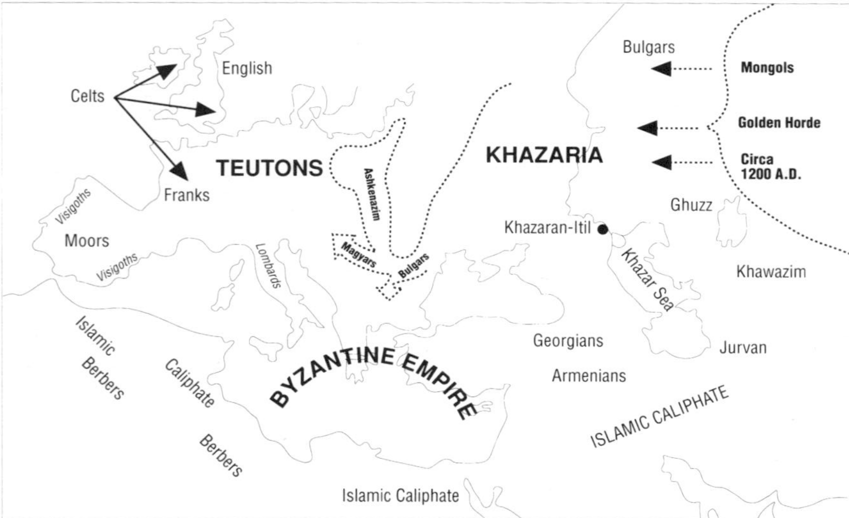

Khazaria and Europe Circa 950-1300 A.D.

Beginning about 740 A.D., "Khazaria was a Hebrew-Cothic speaking empire which became converted to Judaism because of geopolitical convenience. Khazaria reached the apex of its power about 950 A.D.; its capital was Khazaran-Itil on the lower Volga River near the Caspian Sea (known then as "The Khazar Sea"), a town excavated by Soviet archaeologists. As the dotted lines indicate, Khazars were forced westward by invading Mongols (1218 A.D.) and they ended up as the "Ashkenazim Jews" of Eastern Europe, while associated steppe tribes (Bulgars and Magyars) became the Bulgarians and Hungarians of today. At the height of their power, the Khazars were enthusiastic and successful slave traders who captured enemy tribesmen and forest Slavs and sold the captives to both Byzantine Christians and Persian Moslems in such numbers that our modern word for slave derives from the word Slav. It is at least possible that the ingrained "anti-Semitism" of gentile Eastern Europeans results from memories of the Jewish/Khazar slav-trade and the terrors it inspired.

battles speak of 300,000 mounted warriors on each side,[39] but figures of this magnitude are now considered to be gross exaggerations. Arab historians were prone to this. If we take just 10 percent of this figure (30,000 mounted horsemen) and then apply known demographic estimates of the number of non-combatants representing each mounted warrior in the field (wives, aged parents, children, domestic and military artisans, herdsmen and agriculturalists, religious functionaries, civic and military administrators), we come up with a ratio of about 1:15 — that is, one mounted warrior in the field to fifteen non-combatants required to put him there. If Arab battle accounts are divided by ten to reduce exaggeration, we end up with a total of 450,000 Khazars.

However . . .

There are other, more reliable accounts of an Eastern Emperor making a temporary alliance with a Khazar chieftain, usually by promising a daughter in marriage, with the promotional assistance of a portrait of the girl,[40] and acquiring "forty thousand horses" in the process.[41] That strength of cavalry, multiplied by our demographic ratio of non-combatants, yields a total of 600,000 Khazar population—and who can say whether *the* Khazar chieftain with whom such an alliance was negotiated was the *only* Khazar chieftain?

It is really impossible to determine the Khazar population at the height of Khazar power some 200 years after they had converted to Judaism. Certainly, there was a minimum of 500,000 Khazars, most of whom were Jewish; a reasonable maximum population might range as high as 1,500,000 circa 950 A.D., most of whom were Jewish. This is not just a "numbers game." It is quite important because it involves nothing less than the modern, 20th century composition of "Jewry."

The original language of the Khazars seems to have been a form of Gothic, which is to say, a somewhat later form of the "Germanic Branch" of primal Indo-European. The earlier form of the same branch of Indo-European, "Germanic/Celtic" had previously developed in the same Caucasus-steppe region but had migrated westward about 1000 B.C. or before. By the time of Christ, this earlier form of Germanic Indo-European was spoken in modern Germany, Holland, Denmark, and northern France. Celtic had travelled on (earlier) into Britain, Ireland, and Brittany (western France). Gothic was a much later (1000-1500 years later?) development of the same dialect of Indo-European that grew up in the steppes of southern Russia and the Caucasus. It was related to "German" but different in important respects that noted linguists have differentiated and discussed. The Khazars spoke this Gothic on their steppes of southern Russia, and supposedly pockets of Gothic-speaking people still existed in Russian Crimea as late as the 1920s.[42] Khazars, after their conversion to Judaism, spoke either Hebrew laced with Gothic (upper classes) or Gothic laced with Hebrew (lower classes). Both forms of speech were also influenced by Slavonic words (Slavonic, yet another, but reasonably distinct, expression of primal Indo-European) because of the Slavonic-speaking peoples who surrounded the Khazar realm. The result of all this was Yiddish, a language dismissed as a mongrel pidgin of German and Hebrew until the 1920s, but then subjected to some scientific analysis, principally by M. Mieses, a noted Germanic linguist.[43] In his *Die Jiddische Sprache* ("The Yiddish Language," Berlin-Vienna, 1924),

Mieses emphasizes that the Germanic elements in Yiddish are not the early "true German" elements that "good German-speaking" Sephardic Jews picked up as they penetrated into Central Europe from the west but rather an eastern dialect of Gothic.[44] The Gothic-Germanic parts of Yiddish were of much later development of Germanic Indo-European language and, then heavily influenced by Hebrew, resulted in a variation which the incoming Sephardic, good German-speaking immigrant Jews and rabbis could barely understand. As Arthur Koestler quotes, a rabbinical tract from 17th century Poland expresses the pious (and wistful) wish: "May God will that the country be filled with wisdom and that all Jews speak German."[45]

Little can be safely said of Khazar culture. Not much is preserved in either Christian or Islamic sources. But we know that their culture was more advanced than the culture of the woodland Slavonic tribes north of the steppes in "Russia." The Khazars were essentially herdsmen and lived predominantly in tents—which is why primitive *Old Testament* Judaism may have appealed to them—but when history brings them to notice (6th-7th centuries), they were in the process of settling down into agricultural and town life. They had a capital city, Khazaran-Itil, on the Volga just north of where that river empties into the Caspian Sea.[46] It was probably about equally composed of permanent log-built houses and nomadic tents, according to season, but their technology was surprisingly high. Both Arab and Byzantine accounts speak of warriors with *steel* chain mail, and it is known that they minted coinage, so a fairly high technology in metal-working is a fact. They herded sheep, perhaps primarily, but also raised horses; their agriculturalists produced vast amounts of grain and orchard fruits; they even had a substantial fishery on the Caspian. Even today, the Caspian Sea is known, in Arabic, as the *Bahr-al-Khazar*,[47] the "Khazar Sea." Arthur Koestler suggests that the "Jewish " dish of *gefiltefisch*, or some form of stuffed white-fleshed fish, is actually an ancient Khazar dish involving stuffed filets of Caspian sturgeon.[48] One group of Khazars were "maritime," and sailed and fished on the Caspian, but history knows very little of these people.

Aside from the fact that these Jewish Khazars of the 8th-11th century Russian steppes must have been the ancestors of modern Ashkenazim Jewry, they became important for another reason: they introduced the *word* and racial concept of *slavery* into the European world. It was, perhaps, the Khazar adoption of Judaism that gave them not only some necessary religious prestige in competition with their Christian and Islamic neighbors and adversaries, but also, perhaps, gave them some reason for discriminating between themselves as the "Chosen" and others, in surrounding forests and steppes, as "Gentiles" to be exploited. After conversion to Judaism, the Khazars exhibited a cohesive aggression that made them, not just the most prominent steppe people of the area, but *the* dominant people of the region. At this time, the word *slavery* came into the Western world. *Slave* and its derivations come from the word *Slav*;[49] and there can be little doubt that the Jewish Khazars supplied captive Slavs in such quantities to both the Islamic and "Roman" worlds that this ethnographic designation became equated, forever after, with the concept of a "sub-human servant without rights or dignity."

The only Khazar artwork with which Arthur Koestler was able to preface his book, *The Thirteenth Tribe*, happens to be a metallic figurine of a mounted Khazar warrior leading a captive by the hair (with another decapitated head hanging at the warrior's saddle). Our modern word *slave* comes from the Medieval Latin (sclavus) word for *Slav*. Slavs and slaves were once the same thing—or so closely associated that the two designations merged. This can only be explained by the fact that Slavic people were introduced into the Latin and Moslem worlds as abject sources of labor, in such large numbers that the idea of "slav" and "slave" became equated. And, those two words *are still equated in all the European languages of the west.*

The only people with ready access to *Slavs* as captives were Khazars and it is from the time of the Khazar steppe empire that we record the influx of such captives as *slaves*. They were bought by both Christian Byzantines and by Moslem "Arabs," but used for the same purpose: human labor, until death. This was nothing new. Everyone had used human beings in this way from time immemorial. But it is noteworthy that the influx of this labor-material was so strong, *only during Khazar times*, that the human labor-material was named according to its source: Slavs. And, only Khazars, according to our present understanding of Eastern European history, had access to, and ready markets for, such human captives in numbers large enough to equate the word *slave* with *Slav*.

Relatively "modern" Jews (ie. 1900-1990) have often referred to the "inexplicable" and "puzzling" outbreaking of pogroms in Eastern Europe during the period 1880-1900 as an expression of pathological and chronic "anti-Semitism" with no rational explanation whatsoever. Could this outbreak of pogroms have been an age-old fear of Jewish-Khazar slave-raiding that was, at that time (ie. 1880-1900), allowed expression in order to divert complaints against

Czarist abuses? There is, or was, a real and abiding fear of the "Jewish bogey-man that would steal Slavic children" among the non-Jewish peoples of Eastern Europe. Is this fear simply "irrational" and a sick expression of pathological "anti-Semitism," or does it derive from real, and remembered, historical experience?

Then, there is something else to be considered. The almost-certain Khazar/Jewish *slav-trade* was immediately followed by the Mongol invasions. The Khazars were one of the most "easternmost" of the "European" peoples to encounter the Mongols and to retreat before them into the west—toward Russia, the Ukraine, Hungary, Poland, and the Baltic states. Could it not be that much Eastern European "anti-Semitism" results from a genuine and understandable confusion between the two peoples: the Khazar/Jews who had so recently terrorized them with their *slav-trade* and the immediately following Mongols who terrorized them with genocide?[50] This seems more than likely, particularly since some Khazar chieftains succeeded in negotiating brief treaties of utterly necessary survival-convenience with the distantly-related invading Mongol hordes. It may well be that the disparate and desperate peoples of Eastern Europe were not capable of distinguishing between their so-recent Khazar/Jewish tormentors and more-recent Mongol ones, more especially as that distinction was blurred by the steppe tribes themselves. Within this context, Eastern European "anti-Semitism" is not so inexplicable, puzzling and pathological—it is completely understandable. It was an ancient, ingrained and never-to-be-forgotten fear suddenly given room for expression by a decadent Czarist government intent on drawing any discord away from focus upon itself.

Although "limited" in domain by the victory of Duke Sviatoslav (of Kiev) over Khazar armies in A.D. 965,[51] the Khazar Empire continued to exist until it, and much else, was destroyed by the Mongol inundation of southeastern Europe starting in 1218 A.D. Linguistics can trace the path of refugee retreat with some precision. The Khazar, Jewish nation was pushed, first, almost directly due west (carrying the Magyars with them) to create modern Hungary with its large Jewish population. Then, blocked by the Pripet Marshes and by the Dnepr River and (even then) huge Dnepr lakes, they were forced northward to the present site of Gomel (Belo-Russia) where they could continue west again—to radiate into Poland and north into the present Baltic states of Latvia, Estonia and Lithuania. But, all along that tragic way, some lingerers, some simply stubborn ones, and lucky ones survived. Their descendants were the "Russian Jews" of the last century and today. Russian "Ashkenazim," whether they emigrated to America (like Sam Goldwyn's family, Golda Meir's and so many of their Ashkenazim co-religionists of 1850-1920) or whether they stayed in Russia to become the dissidents of the 1960s-1970s. Most of these Mongol-fearing refugees pushed on into Poland, becoming the "Ashkenazim" there. Many emigrated to America, but many stayed to perish in Hitler's holocaust. Some, like Arthur Koestler's family, had lodged early in places like Hungary during that long retreat, and had found a haven, but their descendants (like Koestler himself) emigrated to the West in our 20th century.

The essential point is that these "Ashkenazim," the Jewry of Eastern Europe and the prime victims of Hitler's holocaust, were descendants of Khazar tribespeople. They were *converts* to Judaism. They have no historical or genetic connection with the biblical Jews of Palestine at all. And this was known among themselves, and among the more astute Jewish and non-Jewish historians and linguists, from at least 1700 A.D. It was a well-known, if disguised and ignored fact of Jewish life, from 1850 onward.

Most of what "we" (Gentiles) in North America consider to be *Jewish*, and most of what even the "Ashkenazim" immigrants into North America consider to be *Jewish*—is not Jewish at all! The Hasidic side-locks, the skull-cap (*yarmolka*), the rabbis' fur-trimmed hats and kaftans, *gefiltefische*—none of that is Jewish. Not in the sense that it originated from the people of Abraham in Ancient Palestine. It is *all* Khazar—it reflects traditions of steppe tribesmen who never had anything to do with biblical Hebrews or Judaism.

Our present concern rests with the outlook of these Khazars who adopted Judaism some 1300 years ago and with their descendants, the "Jews" or "Ashkenazim" of today.

They originated in the Caucasus and the steppes to the immediate north—or, at the least, Judaism found them there.[52] And, as such people, they can be presumed to have reflected strongly Neanderthal characteristics. They originated in the area of strongly lingering Neanderthal physical traits. Late Neanderthal sites, discovered by modern archeologists, abound in the old Khazar heartland. It may be no accident, and not only political convenience, that Khazars adopted primitive Judaism. Perhaps *Old Testament* law, and Jehovah's personality, suited their own better than evolved Islam or Christianity.

Not much is known about the physical appearance of Khazars; medieval chroniclers seemed to have assumed reader familiarity with steppe tribespeople and to have left it at that. However, the Khazars are thought to have been one of the component parts of the Huns—of "Attila the Hun" fame—who either remained on the steppes and did not join Attila's celebrated invasion of Europe, or who returned to their homeland once the brief invasion was over.[53] There are some descriptions of Huns, but the goblins described are obviously colored by real terror. Attila, *per se*, was king of the Huns from about 434 A.D. until 453 A.D. and was called "Scourge of God" in his time. He extorted tribute from Rome and was defeated by Aetius at the Battle of Chalons (451 A.D.). The *Columbia Viking Desk Encyclopedia* (page 68) says that "accounts of his savagery, though based on fact, are probably exaggerated."

Huns were a Caucasoid steppe tribe known from China to Europe for their violence and atrocities. They were generally of medium stature, short and stocky, with shortish legs and a rather longish torso, quite hairy and they had big "bestial" heads and faces with large noses. They were usually regarded (by Western Europeans) as being some kind of beast-men, rather than fully human. Their hair could be long and wavy, or short and crinkly, and was brown, reddish brown or black—seldom or never lighter in color to blond. Honoria, sister of a Roman Emperor, offered to marry Attila in order to save the empire, and this was generally considered a fate far worse than death—which gives us an idea of how Huns were viewed at the time. Honoria's gesture was a vain one; Atilla was willing to marry the lady, but demanded half of the Western Empire as dowry, territory he intended to conquer and ravage anyway, and this was unacceptable to the Emperor of the West. Thus, the Battle of Chalons and Attila's defeat.

Although Khazars were a part of the Huns, and a contingent of Attila's entourage, 300 years separate Attila from the first Khazars to emerge into any good light of history. Then, the Khazar Empire lasted for four centuries more before they became the Ashkenazim around 1250-1300 A.D. The Huns obviously exhibited rather pronounced Neanderthal traits, which is not surprising considering their time and Caucasus origin. To what degree these traits were perpetuated in the final Ashkenazim is a matter for controversial debate since we are talking about the vast majority of modern Jewish people. Probably, the Huns' strong Neanderthal traits were much-diluted among the Khazars themselves and then diluted again as the Khazars undertook a long retreat in the company of mixed refugees. Is there, among Jews descended from these Khazars, what can be called "the Jewish appearance?"

Anthropologists, among whom many are Jews, unite in voicing a thunderous *no.* And, respected Jewish writer, Arthur Koestler, quotes all of these statistics in *The Thirteenth Tribe*, Chapter VII "Race and Myth" (pages 181-200); but then he concludes by saying that, anthropological statistics aside, he believes that there is a "Jewish look" along with the common sense beliefs of most non-Jews. Principally, Koestler suggests it has something to do with the nose. It is not just a matter of size, but the formation of the nostrils. Somewhat whimsically Koestler calls this "nostrility" and presents the following argument:

> Thus the nose alone is not a very safe guide to identification. Only a minority—a particular type of Jew—seems to have a convex nose and lots of other ethnic groups also have it. Yet intuition tells one that

the anthropologists' statistics must somehow be wrong. An ingenious way out of this conundrum was suggested by Beddoe and Jacobs, who maintained that the "Jewish nose" need not be really convex in profile, and may yet give the impression of being "hooked" due to a peculiar "tucking up of the wings," an infolding of the nostrils.

To prove his point that it is this 'nostrility' which provides the illusion of beakedness, Jacobs invites his readers to 'write a figure 6 which a long tail (Fig. 1); now remove the turn of the twist, as in Fig. 2, and much of the Jewishness disappears; and it vanishes entirely when we draw the lower continuation horizontally, as in Fig. 3. Ripley, quoting Jacobs, comments: 'Behold the transformation! The Jew has turned Roman beyond a doubt. But what have we proved then? That there is in reality such a phenomenon as a Jewish nose, even though it be differently constituted than our first (anthropological) assumption (the criterion of convexity).'

But is there? Figure 1 could still represent an Italian, or Greek, or Spanish or Armenian, or Red Indian, nose, 'nostrility' included. That it is a Jewish nose, and not a Red Indian, Armenian, etc., nose we deduce—at a glance—from the context of other features, including expression, comportment and dress. It is not a process of logical analysis, but rather in the nature of the psychologist's Gestalt perception, the grasping of a configuration as a whole.

Similar considerations apply to each of the facial features considered to be typically Jewish . . . Taken separately, they are common property of the most varied nations; put together, like an identikit, they combine into a prototype of—to say it once more—*one particular* [emphasis Koestler's] type of Jew, of Eastern European origin, the type with which we are familiar

> ...our identikit would not fit other types of Jews, such as the Sephardim...[54]

Koestler writes "type of Jew, of Eastern European origin, the type with which we are familiar" because of an astonishing statistic. In the 1960s, the number of Sephardim was estimated at about 500,000. At the same period, the number of Ashkenazim was estimated at about 11,000,000. In short, the vast majority of Jews in the world are Ashkenazim *who are descended from Khazar converts to Judaism and who have no genetic or historical connection with biblical Jews or Palestine at all.*

The Jewish "nostrility" which Koestler, quoting Jacobs, is trying to pin down is, quite simply, the remnant Neanderthal nose—that extended "radiator" to warm incoming glacial air—and "Jewish" nostrils can be seen on Carleton Coon's facial reconstructions, based on actual Neanderthal cranial bones, now reposing in the University Museum of the University of Pennsylvania.[55] This type of nose is common among Eastern European Jews, the Ashkenazim, simply because some 1,000,000 or so Khazars fled into Eastern Europe and carried strong Neanderthal, Caucasus, physical traits with them. If it is hard to pin down "Jewish features," and it is, the difficulty is obvious: most European populations preserve diluted Neanderthal physical characteristics, to a greater or lesser degree, depending upon the migration routes of post-glacial times. If "common sense" observation and "intuition" tell us that there is a "Jewish type," it is because we recognize a *concentration of much-less-diluted* Neanderthal physical characteristics in Eastern European Jews or Ashkenazim.

History tells us in no uncertain terms that the Huns

were ferociously aggressive and that the Khazars after them were only a little less so—they were acknowledged to be fierce warriors, they held the regional balance of power, which is precisely why Byzantium and Persia courted them. Their slave trade defined "slave" for all of Europe and they were enthusiastic raiders of people around them (Magyars, Alans, Bulgars, and forest slavs of what would soon be called "Russia"). That being so, the Ashkenazim should also be highly aggressive.

The author submits that they are highly aggressive, and also highly sentimental (the other side of the coin of Neanderthal psycho-biology), but that in Eastern Europe itself this aggression changed its mode of expression. The Jewish Khazars arrived there as refugees; their cohesion as Khazars had been destroyed. They settled in scattered communities, *shtetls*, which were like the villages they had in Khazaria. But they were no longer a unified and coherent nation of warfare, although *shtetl* life must have been primitive, brutal and psycho-sexually violent—much like, say, the Cossack village life described by Mikhail Sholokhov in *Quiet Flows the Don* since much the same people, and lifestyle, are involved. It may be, too, that their non-Jewish refugee neighbors continually persecuted them in retribution for the recently-ended Khazar slav-trade, and so the scattered *shtetl* communities had no opportunity to coalesce themselves into a larger polity. They remained isolated. Then, soon after all these refugees from the Mongols arrived in Eastern Europe, rulers like Casimir III began trying to weld the region into a "modern" medieval European state—replacing the country of wood with stone cities. Once they arrived in Eastern Europe, the Ashkenazim were forever after isolated and surrounded by more powerful groups of non-Jews intent on building a medieval kingdom. Gradually, the Ashkenazim, or some of them, were forced from the rural communities into the growing cities—and also migrated freely into the cities as part of the common urban drift. Because of their shared culture and unique religion, they naturally tended to stick together in the cities. The Eastern European ghettoes were born. In this situation, too, the Ashkenazim were subjected to hostility from their neighbors—possibly because of the still-remembered slav-trade, but probably just as much because of the Ashkenazim clannishness and their strange customs and dress, steppe culture they retained which increasingly set them apart from their assimilated neighbors. Under such isolated and beleaguered circumstances, their natural aggression had to be sublimated into behavior that would not provoke retribution. Their hypothetical Neanderthal psychobiological anti-feminism also had to be sublimated into some sort of behavior that guaranteed female survival or else their people would perish; women, and especially motherhood, had to be pedestalized, protected, when back on the steppes women probably did not have much status among the Khazars—just as women did not have high status among the original Hebrew herdsmen.

Sword-wielding aggression was sublimated into a less dangerous "pushiness" that could be expressed in trade and commerce, a determination to succeed by hard work, and a veiled contempt for their neighbors that was expressed as a wry brand of humor. They clung to Judaism and to their steppe customs and dress as the only identity-giving possessions left to them, jealously preserving them, and naturally also confusing them. Some purely steppe cultural traits came to be regarded as, and be a part of, the Judaic religion as they practiced it: the sidelocks, the rabbinical kaftan and fur-trimmed hat, gefiltefische and Yiddish. Although they retained Judaic discriminations against women, they sentimentalized womanhood and motherhood.

We begin to recognize the "Jewishness" we are familiar with.

VI
Delusions, Deceptions and Distortions

About the time when the Ashkenazim were largely resettled into their Eastern European shtetls (new communities similar to the ones they had had in Khazaria) and about the time they were feeling the first pressures to assimilate into the modernized and westernized feudal states that kings like Casimir III were intent upon molding—about 1400 A.D.—true Jews in Spain, those Sephardim who could actually trace their lineage back to biblical Palestine, were heading into a major crisis.

For the better part of seven centuries, the Sephardim had prospered in Moorish Spain and had achieved great wealth and learning. For the most part they were not just tolerated by their nominal Moslem overlords, but were valued and respected members of the cosmopolitan Moorish community. Many had achieved high positions in the government. This situation was in sharp contrast to their status and security within Christian countries. The problem, about 1400 A.D., was that the Christian "reconquest" of Spain seemed possible in the immediate future. This had begun, in a sense, with the exploits of Rodrigo (or Ruy) Diaz, Count of Bivar, better known to history as "El Cid" (from the Arabic Sayyid, "Lord"). Although he is popularly regarded as a Christian warrior who fought only Moors, El Cid's formidable sword swung on behalf of both the kings of Castile (Christian) and the emir of Saragossa (Moslem). Yet, he is remembered as a Spanish national hero because, he conquered the Moorish stronghold of Valencia (1094 A.D.), which he ruled until his death five years later (1099 A.D.). The capture of Valencia, the first notable victory of Christians over Moors, initiated the idea (especially among Castilian kings) that maybe all of Spain could be retaken. By 1400 A.D., a great deal of Spain had been retaken, including the brilliant Moorish city of Cordova, which fell to Ferdinand III of Castile in 1278 A.D. (It had been the birthplace of Seneca, Lucan, Averoes and Maimonides, a center of culture and learning since Roman times.)

By 1400 A.D., only a few outposts of Moorish rule were left in Spain—and, of course, the incredibly beautiful, strong and populous city-citadel of Granada, the soul of Moorish Spain. Granada looked strong, and was strong—but the handwriting was on the wall. Aside from the reconquest, Spain had another problem: Portugal. Until 1139 A.D., this smaller portion of the Iberian Peninsula had been a part of Spain, more or less, but then broke free into an independent kingdom when Henry, a count of the region, decided to style himself king under the name of Alfonso I. He backed his play with the conquest of Lisbon (1147 A.D.). Under Alfonso IV (1325-1357), who sent a series of Portuguese expeditions to try to find a route to the East Indies by sailing around Africa, Portugal threatened to become Europe's premier power.

Sephardic Jews in Spain and Portugal kept a close eye on these maritime activities. Jews were heavily involved in them. Jews living in regions that Christians had retaken from the Moors sometimes "converted" to Roman Catholicism. The massacres of Jews in 1391 A.D. and 1411 A.D. that swept the Iberian Peninsula influenced their decision to convert. It has been estimated that roughly 100,000 Iberian Jews accepted baptism. They were known as *conversos* or *marranos*, but it was an open secret that the vast majority of them practiced Judaism in private. Indeed, the reality of conversos and marranos was accepted as a polite farce in Iberian society. These Jews were tolerated because they had expertise that the semi-barbaric Iberian Christians lacked. They possessed a heritage of learning matched, perhaps, nowhere else in the world, plus trade and commericial experience and aptitude. Many of them became valued advisors to kings. For a while, as long as they professed credible outward Christianity, they kept their lives and their wealth.

Many of these Jews were involved in maritime-related occupations. Most of the Iberian cartographers were Jewish; at least, the very best of them were, and for a good reason: to compile accurate maps required reading sailors' accounts of voyages in many languages (Arabic, Italian, Scandinavian, English and French as a start) and only Jews were sufficiently literate in the required languages. Also, map-making required the exercise of mathematical and geometric skills which, again, were a legacy of mostly Jewish learning. Thus, a kind of tradition grew up in most European port cities, by no means only in Portugal and Spain, where chart-making and book-selling were merged. Since buyers of charts were sea captains and ship owners, it seemed logical to add ship chandlery to the offering of maps and books. For the non-nautical, a ship's chandler is a sort of ship department store, a one-stop shopping place for everything from ropes needed aboard to specially prepared food, and navigation instruments. And, again, making navigational instruments became a largely Jewish specialty because many had the mathematical know-how while others had been jewellers.[1]

Since Jewish ship chandlers became a natural meeting place for the practical mariners on the one hand, and for shipowners on the other (who paid for what the captains bought), it was only natural that Jewish ship chandlers gradually became agents for ships and fleets: that is, they would try to arrange for cargoes that ships could carry to a given destination since they generally knew which ships were bound for what place. Since some cargo could be very valuable, it naturally followed that Jewish financiers became involved in brokering cargo (ie. buying the cargo, at least on paper, assuming the risk of losing it on the voyage and selling the paper representing the ship's manifest for a percentage of the cargo's worth). They had the capital to purchase cargo, whereas a ship owner might not. It is only a step from that to insuring cargoes and voyages.

Although it is seldom even thought worthy of serious historical research, almost every aspect of European maritime commerce was in Jewish hands and in virtual Jewish control—except the dangerous and brutal business of actually sailing the rat-infested and scurvy-ridden ships on stormy oceans.

Jewish cartography is, indeed, a specialized area of research that is often overlooked. It is said, for example, that the Majorcan master cartographers, Abraham and Judah Cresques (the latter known politely as "Jacomo de Majorca") had maps of every country in the world as early as 1377 A.D., places that Spanish and Portuguese explorers were not to discover until the 1550s. A similar assertion was made by the 13th century Arab cartographer, Abulfeda. There is no reason to doubt these claims, particularly as it is known that Columbus had a map, which he consulted during the 1492 crossing and in accord with which he changed course in order to make his landfall. In *Holy Grail Across the Atlantic* and *The Columbus Conspiracy*, there are detailed discussions of the inexplicable portolan maps and their startling implications.[2]

For now, it is relevant only that Portugal had begun Atlantic explorations as early as 1325. Spain wished to follow this lead, particularly after 1453 when Byzantium finally fell to the Arabs and Moslems gained absolute control of Far Eastern trade. But Spain had a complicated internal "Moorish Problem" to solve before she could emulate Portugal's maritime explorations.

All this came to a crisis of sorts between 1485 and 1492. First, in 1485, the Portuguese finally rounded the Cape of Good Hope and could reach the "Spice Islands" and China, outflanking by sea the Moslem control of land caravan routes. Then, in Spain, Christopher Columbus claimed to have a plan for reaching the Far East by sailing west. In some way still not fully known to history, despite the current spate of Columbus books, the Columbus project was apparently bound up with these events: the bloodless surrender of Granada, the last Moorish stronghold in Spain, in April 1492, the explusion of practicing Jews from Spain, scheduled for August 3, 1492, and the departure of Columbus, also on August 3, 1492. Further, Columbus's voyage was financed by four *marranos*: Luis de Santangel, Alonso de Harana, Alonso de Caravajal and Diego de Harana—four of the richest men in Spain, all former Jews. Further, Columbus's mistress at the time, Beatriz de Harana, was also a converso and related to two of the Columbus backers. She was the mother of Ferdinand, Columbus's second son.

These coincidences have long fed the suspicion that Columbus himself was Jewish, a suspicion first advanced in a scholarly work by Salvador de Madariaga in his *Christopher Columbus* (1967) and re-hashed and sensationalized by famed Nazi-hunter, Simon Weisenthal, in his *Sails of Hope: The Secret Mission of Christopher Columbus* (1981). Although a Jewish Christopher Columbus satisfies some of the facts, it does not explain all of them, and the idea is now generally rejected by scholars (Jew and Gentile alike). *The Columbus Conspiracy* (Houslow, Toronto, 1991) seems to fit all the facts and suggests that Columbus may have been a Cathar heretic.

Whatever the truth about Columbus personally, there is no doubt that Jews and Jewish money were heavily committed to the transatlantic voyage. Even more relevant is the fact that Jews formed a considerable proportion of the "Spanish" settlers and conquistadors who came to the New World. Very possibly, Jews formed the majority of the early

(1494-1520) "Spanish" emigration.

In 1506, the Bishop of Puerto Rico complained to Their Majesties (ie. Ferdinand and Isabella) and the Vatican that ships were bringing mostly Jews as colonists. Four years later, in 1510, the Bishop of Cuba made exactly the same complaint to the same authorities. The official Royal census for 1545 noted that 25 percent of Mexico City's population were openly practicing Jews, and this figure did not include "secret Jews, conversos, heretics and secret Muslims." If this is not an exaggeration, then the bulk of Mexico City's population must have been non-Catholic. And it may not be an exaggeration because in 1527 Francisco Fernandez de Castillo wrote: "There are more Jews than Catholics in Mexico City." Pinto de Lima, writing from South America in 1552, confirmed the same situation there when he wrote that "three-quarters of the Spaniards in South America are of Jewish blood." Three years later, also from South America, Laerte de Ferreira simply commented "Jewish blood is common among the conquistadors."[3] There must be some truth in such statements because, according to official Spanish records, by 1650 there were 15 synagogues in Mexico City alone, three in Puebla, two each in Vera Cruz and Guadalajara, and one each in Zacatecas, Campeche, Merida and Monterey.[4]

The point of all this is that we now come to an issue not presented in any popular history of the Jewish people or any of the numerous books on "anti-Semitism." Not a whisper or reference to it has entered our public school system. And it is, simply, that *Jews were major participants in the genocide of American Indians, as "conquistadors," and were major participants in the greatest human tragedy history knows (no, not Hitler's "holocaust"), the transatlan-*

tic trade in Africans as slaves.

Now, this must have been the case because it was the conquistadors who first murdered and then enslaved the American Indians during the first century of colonialism. Before this first century of exploitation was half over, the supply of Amerindians had already gotten scarce in some places.[5] Then the idea of importing Black Africans in large quantities to replace them was conceived. Indeed, the slave-trade became *the largest single European maritime activity.* Since, on much evidence, Jews were well-represented among the conquistadors, if not an actual majority of the first waves of them, and since Jews virtually controlled the administrative and financial aspects of European maritime commerce, the conclusion is inescapable that Jews must have had a large and profitable role in New World genocide and slavery.

There is no way to estimate the number of victims. Reasonable estimates suggest that from 10-15 million Amerindians and Africans perished in slavery,[6] but some estimates range as high as 100 million victims.[7] European research in this area is, perhaps understandably, scarce. But, ironically enough, one scholarly book written on this subject was by a Jew, Joseph Jacobs, and his book is even more ironically entitled *Jewish Contributions to Western Civilization* (1912). Jacobs chronicles some of the Jewish contributions to a flourishing African slave trade.

In general, the Spanish and Portuguese slave trade was financed out of Seville and Lisbon by financial moguls of which the Harana family (*marranos*) was notable, the same people who backed Columbus. Stephen Birmingham's *The Grandees* provides limited documentation of this "Catholic" slaving financed significantly by Jews. The English,

Dutch and German slavers were financed by Amsterdam Jews (see Jonathan Israel's *European Jewry in the Age of Mercantilism*). In fact, the major center for Dutch slaving was Amsterdam's largest synagogue:[8] finances, insurance, fitting-out and profits were negotiated there. Later, a center of U.S. slave-trade financing was the synagogue in Newport, Rhode Island (the first synagogue in the United States) where Aaron Lopez personally controlled about two dozen American slave ships, while his Jewish colleagues brought the Newport total to about 200.[9]

This is not to suggest that the Jews were solely responsible for the initiation and maintenance of the transatlantic slave trade—or the *Maafa* ("Holocaust"), as some African Americans term it—but it is certain that Sephardic Jews were major players in it and profited hugely from it. Jews, along with non-Jewish Europeans, must bear their share of the blame and guilt of this greatest humanly-engineered tragedy of history. Yet, you will not find any reference to Jewish participation in the slave trade in our history books. And, when Dr. Leonard Jeffries, Chairman of the Department of Africana Studies at the City College of New York, referred to it in a speech on July 20, 1991, the Jewish-controlled Eastern Media Establishment attacked him as being "anti-Semitic." *The New York Times'* Metropolitan News section (August 7, 1991) splashed a full-page headline: "City College Professor Assailed for Remarks on Jews." A Jewish colleague of Jeffries at City College, Dr. Bernard Somer, could not believe the attack: "Len, everybody knows the Jews helped finance the slave trade," he said, and advised Jeffries to take the matter lightly.[10]

But this was no light matter, as Jeffries well knew. Perhaps historians have known that Jews had helped to finance the slave trade, but that is not "everybody"—and, in fact, "everybody" (that is, the general public) up to the educational level of high school and beyond has been carefully shielded from this information. It could not be denied that the English, Dutch, French, Americans, Spanish and Portuguese (not to mention the Danes, Germans and Arabs) were involved in the slave trade—but that the Jews had been financially involved with all slaving nations has been suppressed. Jeffries was subjected to what he called a "media lynching."[11] He dared to break the unspoken rule: if Jews are mentioned at all, it must be as innocent victims, never as aggressors. All novelists know that, all script writers and TV/film producers know it. They also know that if they break that rule, their novels just do not get published, their scripts do not get produced and they can forget a TV/film production career in Hollywood. Obviously, Jeffries was well aware of the rule. He was just someone who got fed up with dancing to the tune of the Chosen People.

One tune. There are many others.

It is worthwhile to discuss the issue of Dr. Leonard Jeffries in a bit more detail because his confrontation with the media poignantly encapsulates major aspects of 3,500 years of Jewish-African inter-relations. Simultaneously, it hints prophetically at future developments in Western Civilization.

Because of the publicity, Edward I. Koch, then Mayor of New York City, wanted to know exactly what Jeffries was saying and what documentation he had to back it up. Koch invited Jeffries to meet with him and his legal counsel, Dan Wolf. But Jeffries consistently refused to go on Koch's turf unless Koch gave him back his pyramid. "Well, I'll come down to your office, but first you've got to give up my pyramid," Jeffries said in a phone conversation.[12]

"What?" Koch asked, baffled.

"I would not even consider a dialogue unless you're prepared to give up my pyramid," Jeffries repeated. "We have a picture of you riding on a camel around the pyramids (at Giza-*Author*) several years ago talking about how your ancestors built them thousands of years ago. We got a picture of Menachem Begin dancing in front of the pyramids that his ancestors built thousands of years ago. We want our pyramids back, and there's no need even to dialogue unless you're prepared to give them up."[13]

In 1977, Menachem Begin visited Egypt as part of the Egyptian-Israeli peace initiative then taking place. Begin visited Cairo to reciprocate Sadat's earlier visit to Jerusalem. And, while in Egypt, Begin did take the opportunity to see the pyramids at Giza, he did cavort in front of them, and was photographed saying that his people had built them thousands of years ago. This photo was widely published in U.S. newspapers and it was a subtle propaganda ploy to undermine Egyptian prestige during the peace talks. The inference was, of course, that the Jews had been responsible for the symbols of Egypt, the pyramids, and thus (in the mind of the average person) for Egyptian civilization. It was a snide and malicious suggestion that the Jews had now regained something that was theirs, that Egypt was being forced to relinquish something—that, in short, Egypt was "giving in" instead of being an equal partner in the 1977 peace initiative. And, just in case this photo of Begin at the pyramids was not enough to propagandize the American public to the Israeli side, it just happened that Menachem Begin's (somewhat rosy) memoirs, *White Nights,* was published and mass-marketed during the peace talks.

Presumably, when Menachem Begin said that his people had built the pyramids thousands of years ago, he was referring to the biblical Hebrews and their well-known servitude in Egypt until Moses said, "Let my people go." But, the fact of the matter is that the pyramids were built about 2600 B.C.[14] Biblical Hebrews did not even arrive in Palestine until 2000 B.C. They did not plead to be let into Egypt so that they could survive a famine and agree to work for their keep until about 1600 B.C. When the Hebrews entered Egypt, the pyramids had been standing for at least 900 years, probably more. The Jews may have worked to build temples, causeways, dykes along the Nile, boats—any number of things—*but they did not build pyramids.* That is just one of the historical distortions of Hollywood epics based on the Exodus theme.

Then, there is the small matter of Menachem Begin talking about *his* people. Menachem Begin was born in Poland and emigrated to Palestine after World War II. His people were the Ashkenazim. They may have made bagels,[15] excellent carts[16] and gefiltefische—*but they did not make any pyramids.*

The Egyptians made them. Africans made them. As a person of African descent, Jeffries, it seems, was being charitably humorous when he said to Koch: "I won't talk until you give me back my pyramids." What Jeffries was approaching, with some degree of civilized humor, was an insidious and malicious distortion of history.

When they finally sat down to talk (Koch, on a trip, had called to say: "Tell Jeffries he can have his pyramids" and so a meeting was arranged), Jeffries tried to educate Koch and Wolf about the Statue of Liberty, but without much success. Koch had a model of it in his office and, like most Americans, thought it had been intended to welcome immigrants to the land of liberty. It was not. The man who conceived the idea for the statue, in 1865, Frenchman Edward Rene Lefebvre de Laboulaye, was a political scientist, a political leader and parliamentarian, author of a three-volume history of the United States—and, in 1865, he was president of the French anti-slave society.[17] With the end of the Civil War and the true end of U.S. slavery, the statue was conceived to celebrate that event. It was, in fact, a rather backhanded gesture of congratulations to the United States for being the last major Western country to abolish slavery.[18] Original artistic studies for this planned statue feature a Black woman bursting the bonds of slavery, with the broken shackles at her feet.[19] As Dr. Jeffries ascertained, several of these models are in the New York City Museum (5th Avenue and 106th Street) *but they are locked away from public viewing.*[20] The French Cultural Center (5th Avenue and 82nd street) has a special "Liberty" edition of the magazine *France* in which the real story is told and some of the models are illustrated. The original concept was not acceptable, even as a gift from France, and the idea was finally modified into a properly Caucasian personification of "Liberty" before the U.S. would accept delivery. Even then, Americans did not seem too keen on "Liberty" because the French people had paid for the completed statue before the Americans even managed to pay for its pedestal. In the final analysis, the very symbol of America has been distorted by Caucasians, just as the very symbol of Egypt was the object of attempted theft by Menachem Begin.

Hollywood, according to Jeffries, has played a prime

role in distorting history, society and keeping the American public mesmerized. On the one hand, the public has accepted Hollywood stereotypes without hesitation, and this has retarded maturity and social change within the United States. On the other hand, this somnolent public has tolerated incessant Jewish and Zionist propaganda in films and on television, which has actually affected U.S. foreign policy in significant ways. Jeffries pointed out that Hollywood, as we know it, was basically the creation of Russian Jews. Again, he was criticized for this statement as being "anti-Semitic," which is absurd: Jews themselves take credit for the Russian Jewish creation of Hollywood [see *Russian Jewish Refugees in America*, (Louis E. Levy, Revisionist Press, 1979), or read *An Empire Of Their Own: How the Jews Invented Hollywood,* (Neal Gabler, Doubleday, 1989)]. Goldwyn (originally Goldfish), Mayer—MGM—Zukor, Fox and Sarnoff (NBC) were all Russian Jews, and there were many more in media who were not so famous. This is fact, not "anti-Semitism." And it is fact that Jews today virtually control—through initial creativity, production, casting, financing and distribution—what Americans see on TV and at the movies. Jeffries did not like what he saw. Only a couple of generations ago, Blacks (according to Jeffries) were stereotyped as "Stepin Fetchits." Today they are presented as junkies and criminals, while Native Americans (until recently) were wiped out regularly every Saturday matinee. Hollywood has been a significant molder of American values and attitudes, and the message has been clear: discrimination against Blacks and Native Americans, and their physical and psychological destruction is acceptable. That was Jeffries' complaint. Is it "anti-Semitic" or a statement of fact?

While denigrating or degrading other peoples and cultures, Hollywood has steadfastly been pro-Jewish and pro-Israel. Seemingly endless films like *Exodus*, *QB VII* and films about "the holocaust" are produced continually—as well as television series like *Sisters* and *Evergreen*. But remember the furor when Vanessa Redgrave wanted to make a film from the Palestinian point of view? In other instances, there is blatant anti-Arab propaganda, often disguised as harmless humor. For example, in Mel Brooks' *Blazing Saddles*, two Arabs on camels show up at a gunfighters' conclave in the Wild West and are introduced as "the most vicious gunmen in the West." We are supposed to laugh, of course, at the anomaly of Arabs and camels in America's West, but the propaganda message is clear. As for "ancient Egypt," the various "let my people go" epics have universally portrayed the Egyptians as being cruel overseers with the Hebrews, under the lash, slaving to build pyramids. The biblical truth of the matter is that the Israelites begged to come into Egypt in order to avoid starvation, agreed to work for their keep, but did not build pyramids. Indeed, in the beginning (at least), Pharaoh offered the Israelites a good deal: *Genesis* 47: 24-27 . . .

> And it shall come to pass in the increase, that ye shall give the fifth part unto Pharaoh, and four parts shall be your own, for the seed of the field, for your food, and for food for your little ones.
>
> And they said, Thou hast saved our lives: let us find grace in the sight of thy Lord, and we will be Pharaoh's servants.
>
> And Israel dwelt in the land of Egypt, in the country of

> Goshen; and they had possessions therein, and grew, and multiplied exceedingly.

So, by the Hebrews' own account, Egypt treated them well (at least, at first). It may well be that the Egyptians treated them with increasing brutality, as the *Bible* says, but the Egyptians maintained that the Jews did not keep their part of the bargain and were continually worshipping their God instead of working. We cannot know the truth of the matter. There may have been injustices on both sides. However, what is certain is that in the biblical account there is not one Jewish or Egyptian reference to making pyramids! The only product mentioned is adobe brick, which requires straw (which the Jews had to gather), and bricks were used for ordinary houses and mundane buildings. Pyramids and temples were made of huge blocks of stone. Whatever the truth of the matter, it seems that a thousand years later, in the time of Jeremiah (roughly 600 B.C.) when the Jews were again threatened by famine and invasion, many of them fled to Egypt (*Jeremiah* 43-46). It may be true that in the time of Moses and the Exodus, Egyptian brutality greatly increased. Or it may be that the Jews simply wanted to leave because general agricultural conditions had improved and they no longer needed Egyptian aid. Also, it may be true that the Egyptians did not want to lose Hebrew labor. But the brutality of Hollywood epics seems exaggerated. Furthermore, in the Hollywood versions of the biblical account (faithful enough), the Red Sea is always shown parting for Moses and the Israelites and then engulfing the pursuing Egyptians. But the *Old Testament* sharply disagrees with the rabbinical traditions preserved in the exegetical works; several commentators emphasize that a majority of the Israelite host was destroyed as well (Rashi, Zohar). There was some sort of tidal catastrophe, and both Egyptians and Israelites suffered heavily in it.

In Hollywood, the Africans (whether ancient Egyptians or more modern Africans) have not been granted the sympathy, or the level of culture, that real history insists upon.[21] Hollywood's portrayal of other peoples and cultures has been, at best, biased and "Westrocentric" and at worst (usually when it comes to Arabs), Hollywood has been obviously biased and blatantly propagandistic.

Aside from Hollywood which produces entertainment, and aside from the chronic Arab-Israeli mess as presented in Exodus-type movies, one wonders to what extent pro-Jewish bias in the news media has influenced U.S. foreign policy. The Gulf War and "Desert Storm" may have been a greater propaganda victory than a military one. Just a year after the great victory and still in a patriotic fervor, Americans may not be too willing to consider that they may have been duped, that Iraq was a fall-guy.

Originally, Iraq charged Kuwait with oil theft: the allegation was that Kuwait was drilling wells obliquely beneath the border. On the surface, the wells were in Kuwait, but far below they were tapping into oil pools on the Iraqi side of the border. Iraq wanted to put a stop to this. Apparently, there seemed to be some truth in these Iraqi accusations—at least, they were given some credence by the U.S. Ambassador at the time. Although the real truth may never come out, it seems at least possible that Iraq obtained U.S. assurance that there would be no massive retaliation if Iraq invaded Kuwait to put a stop to the "oblique drilling" technique. So, Iraq proceeded.

Now, this is not to say that Iraq's leader is a nice guy.

And it is not to deny that Iraq had built up (on paper) what was supposedly the fourth largest military machine in the world. And it is not to deny that this military machine was really, ultimately, intended for wiping out Israel (if Iran could be held at bay). Now, after the Iraqi invasion of Kuwait, the U.S. public was not inclined to support massive U.S. military intervention in the Persian Gulf. Kuwait spent $6 million on market research and a subsequent promotional campaign to find out what would inspire the U.S. public to back a war. They discovered that the American public was moved by atrocities, particularly atrocities committed on babies. Very quickly, then, at congressional hearings, the televised scene of an attractive young Kuwaiti woman who swears she witnessed that very sort of atrocity was presented. According to her, Iraqi soldiers burst into a Kuwaiti maternity hospital and snatched approximately 100 newborn babies from incubators and flung them on the floor to die. The young woman was weeping emotionally. This story of the "incubator atrocity" was used in many subsequent speeches, including those by President Bush, to incite the U.S. public to demand military action against Iraq. What followed was the televised brilliant and exciting "Desert Storm" spectacle.

But now, a year later, it transpires that this young lady was the Kuwaiti Ambassador's daughter. She may not have even been in Kuwait when this alleged atrocity took place. Doctors and nurses at this hospital remember no such atrocity. In short, a scheme was concocted to get America into the Persian Gulf war; a scheme based on careful psychological research and a scripted atrocity to match. Doubtless, there were atrocities. There always are in warfare—but there did not seem to be the atrocity that was so carefully researched and recounted to affect the American psyche. And, it must be admitted, Sadam Hussein made an excellent villain, is an excellent villain.

Yet, the Gulf War seems almost too convenient. Consider who wins:

First, the United States gets a walk-over victory against Iraq. The U.S. is in need of a military victory. There was the debacle in Vietnam, the inconclusive and scandal-ridden operation in Costa Rica. The U.S. could not even rescue its own hostages from Iran (helicopters broke down), and only the quick-thinking Canadian Ambassador to Iran, Ken Taylor, was able to get a few Americans out. The U.S. badly needed a military operation that was a success, not only to restore credibility to the American military but also to drum up some patriotic, martial fervor. Iraq was the perfect fall guy for this. Strong on paper, weak in reality. If the United States could not beat Iraq, then the West is in real trouble.

Second, with Kuwait and Saudi Arabia suddenly Arab allies, and allies with oil, American's petroleum requirements seem assured for the immediate future. America has some credibility with some Arab states—and these states are far enough from Palestine to be no threat to Israel.

Third, Israel. Iraq's formidable military machine is destroyed—a real threat to a country as small as Israel—yet Israel does not have to lift a finger to do it. Israel is not the aggressor in the eyes of the world.

Some of the facts around the initial Kuwaiti "atrocity deception" have already been revealed by the CBS *20/20* television program.[22] Most likely more and more evidence of collusion and collaboration will come out regarding this too-convenient war; and probably part of the deal was cut in Jerusalem. Israel was the big winner in the Gulf War, so simple logic compels that conclusion.

If the evidence begins to accumulate that way, then the Gulf War must be seen as something made possible by the American public's willingness to be duped when it comes to Arabs. This willingness was created through a long process of careful nurturing by Hollywood and the Eastern Media Establishment.

Dr. Leonard Jeffries has his own suspicions about the Gulf War or so it seems from some rather cryptic comments published in *Emerge*.[23] However, even if the Gulf War was completely legitimate, it nonetheless illuminates the disparate, unequal and disrespectful orientation of American society toward African people. Many, if not most, of America's combat troops in that war were Black. Heroes last year, a shocking number of them are presently destitute and living on the street.[24] African Americans seem always to have been expendable pawns in the Western game, not the only pawns, but the majority of them.

However, this is changing. Only a generation ago, the majority of African Americans seemed to want nothing more than assimilation. Sometimes, despite the economic and social inequalities and physical abuse, they were nonetheless incredibly patriotic and courageous. Black units in World War II served with exemplary distinction, and the "Tuskegee Airmen" achieved a record in bomber protection that may never have been equalled by any fighter pilots, of comparable numbers, in the world. Yet, they lived beside (not among) and fought and died for white Americans who "treated them like dirt."[25] And, in due course, by such sacrifice and almost inhuman patience, some African Americans have finally achieved a measure of security, "integration" and "assimilation." From this foundation, many of them are looking at the Western, Caucasoid society their parents strove so hard to be a part of and are asking the unspeakable question: Do I really want it? And the answer for many is *no*. It is not a harmonious society, it reflects few truly humane values, it is destructive of others and itself, and it is terribly violent militarily and psychologically. Emerging Black intellect is increasingly looking toward people from the Caucasus with an objective, critical eye—not at all an envious one.

In my view, things have reached the following stage of conceptual evolution: African Americans (as well Asian Americans and Native Americans) have long struggled against what might be called the superficial layers of the Western World's cultural onion. They have achieved social victories of equality and dignity one by one. Although there was much ignorance, hatred and discrimination to overcome, there was also much assistance and encouragement, so long as the "visible minority's "goal was to assimilate within Western culture, American culture. And assimilation, by definition, means to accept and adopt cultural values, myths, attitudes and behavior. But, as these non-Westerners work toward the center, or core, of the onion, they discover two things.

First, they simply cannot accept the values, myths and attitudes of Western society because these aspects of Western civilization are based on a (to them) peculiar and puzzling Caucasian psychology, which maintains itself by distorting history and purporting human values shared by no other culture. An ever-increasing number do not want to assimilate into it. Rather, they have begun to look toward their own cultural and racial origins for more sensible and more human values than they see in the West. At present in America, this is happening mostly among African Americans, but is only a matter of time before Asian Americans and Native Americans begin seriously to do the same thing. (Dr. Jeffries is one of many African American academicians who has learned enough to see Western Civilization for what it is.)

Furthermore, these Black intellectuals have a second realization as they work toward the core of Western values: that the core of Western culture—that concentration of Caucasus psychology—is essentially Jewish. At least, that core has been molded and conceived to an inordinate degree by Jews. The social evolution of the West is at the stage that Black intellect (not wanting to become Westernized) is now beginning to confront the core, the concentration, of what Western Man is all about. We will witness a steadily escalating confrontation between Black Intellect and Jewish Intellect, African souls and Caucasus souls.

The educated and inquiring "visible minorities," of which African Americans are only the most numerous group in North America, have peeled away enough layers so that they have glimpsed, at last, the image of a glacial god that permeates all of the West to a greater or lesser degree.

All of the West exhibits "the iceman inheritance," but non-Whites are starting to grasp that the psychological core of it originated among people who emerged relatively recently from the "migration backwater," that Neanderthal refuge of the Caucasus-Elburz. This region has spawned human populations of unique religious chauvinism, sentimentality, violence, intolerance, cruelty—perhaps more so than other places in the Caucasian domain. Much of this region falls into what we call the "Middle East," and the area has spawned crises from biblical times until today.

I do not think that any one can objectively argue that the Jews are, inherently, any more Neanderthal than other peoples originating in this area. But I think that something can be said: it was Jewish fate to have evolved a religion and a culture at a time, many thousands of years ago, when the Neanderthal characteristics were more pronounced than they are today. That religion and culture, to some degree, perpetuates attitudes and values of Neanderthal-like intolerance, identity-hunger and psychosexual discriminations that are no longer really so impelled by much-diluted Neanderthal traits. Other peoples, though originating from the same region and also aggressive, emerged later and without a religion and culture dating from a much earlier time. One might say that Jewish aggression is not greater than other groups from the same geographic region but more culturally and religiously codified and perpetuated. We see, for example, after more than 4000 years, a perpetuated Jewish belief in actually being a Chosen People. That is a quite primitive orientation, one anachronistically surviving from a long-vanished tribal and parochial world (an orientation perpetuated by a religion molded in the distant past).

That idea of being the Chosen People and having been promised Palestine by their very parochial God has fostered several delusions and distortions upon Jews and everyone else. The ancient Hebrew urge for identity is so strongly entrenched by religion and culture that it has demanded preservation by intellectual deception.

The existence of the modern state of Israel is a case in point. We have seen that the claim of biblical Jews to a "homeland" in Palestine is shaky; their claim is no better than a similar claim by many other people who have inhabited the area. But, when we now consider that 95 percent of the Jews in the modern world are Ashkenazim, their claim to a Jewish homeland in Palestine becomes absurd, grotesque. They are descended from Caucasus converts to Judaism, nothing more, and have no genetic or historical connection with Palestine at all. The case of Zionism is no different from, say, a group of Cree Indians from northern Canada becoming converted to Catholicism and then demanding a piece of the Vatican as a "Cree homeland."

Israel was conceived by Ashkenazi Jews to buttress their delusions, to feed that insatiable desire for identity-in-time. Israel was established by deceiving the rest of the world, but particularly the North American public, about the real origins of the Ashkenazim. That deception was made possible by Jewish influence in U.S. mass media and international finance. Arthur Koestler, a celebrated Jewish author, was not unaware of the anomalous situation of modern Israel in the light of Ashkenazim origins. Yet, in *The Thirteenth Tribe*, Koestler falls back upon a legalism, a technicality, to defend Israel's right to exist:

> But that right (ie. of the State of Israel to exist) is not based on the hypothetical origins of the Jewish people, nor on the mythological covenant of Abraham with God; it is based on international law—ie. on the United Nations' decision in 1947 to partition Palestine, once a Turkish province, then a British Mandated Territory, into an Arab and a Jewish state. Whatever the Israeli citizens' racial origins, and whatever illusions they entertain about them, their State exists de jure and de facto, and cannot be undone, except

by genocide.[26]

That is all very true, but what Koestler neglects to say, or admit, is that the United Nations' decision to create Israel resulted from a deliberate deception made possible by propaganda and money. In a court of law, which Koestler would seemingly invoke, the 1947 contract was negotiated and executed under false pretenses by one party, while the other party (Arabs, Palestinians) was not really heard at all. The United States, which largely initiated and funded the United Nations, possessed a population thoroughly propagandized by Hollywood.

Nonetheless, Koestler was correct. Israel exists, "and it can not be undone except by genocide." Whereas that is an unacceptable solution for humanitarian reasons, it is probably not even possible in a practical sense. It is an "open secret" that Israel possesses a formidable nuclear capability. It is not so well known that there's "the Masada Plan": a one way nuclear strike by an elite wing of the Israeli Air Force, to destroy every Middle Eastern capital and major city possible if it looks as if Israel may be defeated.[27] It is a kamikazi-style strike by the handpicked pilots involved, for there would be nothing worth returning to. If the Neanderthal hunger for identity-in-time is as strong as it seems, then Jews in Israel would prefer to bring about a nuclear holocaust than lose Israel. At least, even in a horror of world destruction, they would have been around at the beginning and the end; it is the ultimate in identity-in-time.

Our limited purpose is not to deal with the confrontation in Palestine nor to advocate any solution. Perhaps the Arab-Israeli confrontation is not a problem (to which there's a solution), but a predicament which does not necessarily have a solution. One can only respond to a predicament in the way that seems most productive at any given time. But, at least, we can try to understand why the Arab-Israeli predicament developed in our modern Middle East, and this book goes a long way toward suggesting a plausible explanation.

The world is presently facing many crises (not just the Arab-Israeli one) and, if we look at these crises with total objectivity, we will see that *all* of them, at base, derive from Caucasoid behavior, Caucasoid values and Caucasoid psychology. Nuclear war, environmental pollution, resource wastage—all have reached a critical mass and have become threats to our survival because of Caucasoid evolution. Within this truth, one can say with validity that our crises are somewhat more acute and developed a bit more quickly than they might have because of disproportionate Jewish influence in Western affairs.

We really have but one hope for survival: *human intelligence*. As we have seen, even Caucasoids possess this quality in very adequate quantity, even if their expression of it tends to be logical to the exclusion of human values. It is there. It is powerful. But, it cannot be exercised for the benefit of human survival if that intelligence has insufficient facts with which to work, or if that intelligence is compromised by exposure only to selected facts—facts acceptable to the personality behind the intelligence.

If the intelligence is to function properly in the interests of human survival, it must be exposed to facts which it might rather avoid. Therefore, our only hope is education of Caucasoid intelligence. We must deluge that intelligence with facts that are distasteful to it so that it can grasp the personality lurking behind logic, controlling and distorting logic for the sake of that Iceman personality's primitive identity hunger.

Not so long ago, it was simply impossible to present distasteful facts for the consideration of Western intellect. Westerners unconsciously censored such facts and consented to deal only with data that was acceptable to them, flattering to them. But that situation is changing now and mostly because of the emergence of an independent Black intellect. That intellect is fueled by scholars and thinkers, expressed by Black-owned publishing houses and mass media. For the first time, there is in the Western World, and in the United States, a *perspective* that is not essentially Western. It is independent and non-Western because the evolution of African Americans is away from assimilation into the Western World. Although this emerging Black intellect is still in its infancy, and may also be prone to its own distortions and delusions,[28] it nonetheless may be the salvation of humanity in the crisis years ahead. It can present to the Western World and to the Neanderthal-Caucasoid "Iceman's" personality another, different and not-so-flattering reflection, composed of facts that have long been ignored.

This is the educational process so urgently needed. Prejudice should not be involved—just the calm and patient statement of unpleasant facts, over and over, until these facts come to be considered by Western humanity.

The Iceman Inheritance is not "anti-White," and *Chosen People From The Caucasus* is not "anti-Semitic." Both books exist as a marshalling of facts. Granted, the facts are unfamiliar to most Westerners and Jews, and they may seem unacceptable. But, if they are facts which seriously affect behavior, they must be dealt with. The purpose of both books is simply to expose certain facts, and a reasonable interpretation of them, in the hope that we Westerners and Jews may see ourselves in a more objective light. Only in this way can we modify our own behavior in the direction of human survival.

To travel in that direction could be an easy journey because it actually involves a reduction of burdens. For Caucasoids, "white men," Westerners, it means laying down that baggage of delusion that we brought culture and civilization to the other peoples of the world, that delusion that has been so aptly called "The White Man's Burden." For Jews, at the core of the Western world's conceptions and self-conceptions, to travel in the direction of human survival means more relief yet, a rejection of even heavier baggage than most Westerners have been carrying. Jews can jettison the terrible obligations and sacrifices of being the Chosen People of an ancient, glacial God. To join the rest of humanity on an evolutionary course of survival should be a relief to Western man, a casting aside of burdens of the soul.

And yet, of course, these burdens are more precious to us than any rich man's bag of gold, despite their crushing weight. They are nothing less than the essence of our Neanderthal identity-hunger, and yet, both their substance and their value are delusions formed by our own unique and traumatic Ice Age experience. *Can* we lay these burdens down? *Can* we exchange delusions of identity for the relief of being only human? This is our challenge as we move into the 21st century.

AFTERWORD

As this book goes to press (April 30, 1992), North American newspapers carry banner headlines about the riots in Los Angeles resulting from the verdict in the Rodney King affair. (Rodney King was the Black L.A. motorist who was brutally beaten by four white policemen on March 3, 1991.) Unlike many similar incidents in the United States (and Canada), the torment of Rodney King was not just another "allegation" lacking objective and dramatic proof; the police brutality was videotaped by an amateur cameraman and the tape was broadcast throughout the world. From the evidence of this tape, prosecutors in the trial of the accused policemen counted 56 blows and kicks delivered by the four white policemen on a helpless and writhing Rodney King during the course of an 81-second beating that was obviously, to any objective person, "excessive use of force." Indeed, anyone watching that videotape—and it required a strong stomach to view it completely—had to agree that the four white policemen were simply indulging in a spasm of pure racial hatred and violence.

Because it had been videotaped, however, this particular incident could not be ignored and covered up. The four white policemen were charged with multiple counts of brutality and brought to trial. Perhaps inevitably, the whitewash mechanisms began to operate almost immediately; *there was not one Black on the jury* even though, given the demographics of Los Angeles, three or four Blacks would have been fair racial representation. Presumably, it was considered that Blacks would be biased against the white police defendants. An Asian-American and a Hispanic-American were the minorities who could be easily intimidated by the remaining 10 white members of this jury.

Nonetheless, even with the racially skewed jury, many North Americans thought that justice would prevail *this* time if only because the evidence was so obvious and unarguable.

Incredibly, the jury acquitted the four policemen on charges of using "excessive force" (definitely an euphemism in this case). Even the state prosecutors were amazed at the jury's verdict. Black community leaders were, perhaps, not so amazed: "If something in you can die, that something died," said Rev. Cecil Murray of the African Methodist Episcopal Church in South Central Los Angeles. Watching the news coverage of the verdict on television in his church packed with fellow viewers, Murray fought back tears and declared the verdict a tragedy—"Not because it's unbelievable but because it is believable." In fairness, it should be added that many white residents of L.A. were equally outraged. One of them, Rose Brown, who drove to the courthouse to hear the expected guilty verdict was interviewed by the *Los Angeles Times:* "I'm not only shocked," she said, "but I'm hurt for Americans as a people. I don't think Rodney King was on trial but America was on trial."

But other whites, and probably a secret majority of them, applauded the acquittal of the four cops. "I'm glad they got off," Barbara Williams told the *Los Angeles Times*. "They did what they were trained to do."

Incredibly, Black community leaders in Los Angeles pleaded for calm and non-violence. But it was a vain appeal, understandably. Rage boiled over at the patent injustice of the so-called justice system, and infuriated Blacks took to the streets to dish out to the white motorists a dose of what Rodney King had taken. Looters ravaged stores, motorists were dragged from their cars and pummeled and demonstrators rushed local police headquarters. About 60 people were injured seriously; a state of emergency was declared by Mayor Tom Bradley; the National Guard was deployed by Governor Pete Wilson. Such is the human tragedy when justice is perverted and denied. The ongoing cost will be that Rodney King will not be forgotten for a long, long time and, if I were a white cop or just a white resident of L.A., I would be very nervous for a long, long time. Justifiably nervous. Guiltily nervous. When justice fails, vengeance is the only, and inevitable, recourse of the disenfranchised.

I do not wish to tie this book to the affair of Rodney King. This entire incident is just one more in a long and depressing chronicle of racism and injustice perpetuated by the Western World and so-called "Western civilization" upon non-whites. It is just one more in a series of *judicial* injustices committed against African Americans since the belated Emancipation Proclamation. The technical, judicial injustice applied to the Rodney King issue is a lingering shadow of 550 years of believable inhumanity practiced upon Africans by the Western Civilization of Europeans and North Americans. Before the L.A. verdict of April 29, 1992, there had been 425 years of absolute slavery of a savagery unknown elsewhere in the world, followed by a century of semi-serfdom and racial denigration after the official end of slavery in America, continued by almost three more decades of racial discrimination in housing, education, employment, health care. The rebellions in Los Angeles which followed the jury's incredible acquittal did not result only from this particular legal case. The *judicial* injustice was the last straw in five and a half centuries of obscene inhumanity. The race riots that shocked America were a natural and inevitable response to the technical and judicial injustice thought (hopefully) to be ended by a videotape so that true *social justice* could begin. But the White Establishment sought, once more, to peg its hopes on a technical and judicial injustice. As far as American "Mainstream/Middle Class/White" society is concerned, well, it may go down in history as the *last* racial injustice they enjoyed.

Consider, for example, that every single juror *swore on the Bible* to be impartial in the pursuit of justice based solely upon the evidence. Each juror, in effect, made a covenant with God to mete out justice based on the truth. This covenant has disturbing implications for non-whites all over the world, *particularly for Blacks in North America,* because it involves a terrible historical irony.

Traditionally, African society was so remarkably non-violent and crime-free that there was not even a (West) African word for "jail." Indeed, this peace and general harmony of African societies amazed and awed early Euro-Arabic travellers south of the Sahara. Ibn Batuta, an Arab explorer, first published his *Travels in Asia and Africa* in Morocco in 1356. About Black African countries, he wrote the following:

> They are seldom unjust and have a greater abhorrence of injustice than any other people. Their sultan shows no mercy to anyone who is found guilty of the least act of it. There is complete security in the country. Neither traveller nor inhabitant in it has anything to fear from robbers or men of violence. They do not confiscate the property of any white man who dies in their country, even if it be uncounted wealth. On the contrary, they give it into the charge of some trustworthy person among the whites, if such can be found, until the rightful heir takes possession of it. (Taken from Basil Davidson's *Lost Cities of Africa*, pg. 79, translated into English by H.A.R. Gibb, 1929)

Ibn Batuta's account may be somewhat exaggerated and romanticized, and if so, it is probably due to his cultural shock at experiencing the African concept of social harmony. Ibn Batuta had travelled widely throughout the known world accessible to Arabs of his day, from Morocco, throughout the Middle East and Central Asia, and on to China and Indo-China. He had a broad grasp of various peoples and cultures of his time from 20 years of first hand experience. If he wrote about the Africans in this way, there must have been more that a grain of truth to it. Elsewhere he states (as did other early travellers) that there was no (repeat, *no*) racial discrimination among the Black Africans.

The Black African concept of the way things should be was a harmony of past-present-future life welded together wholistically by a compassionate religion and social structure (as human cultures go). When Black Africans were torn from their traditional cultural and religious fabric by the European slave trade of 1444-1820, and when unfortunates from various parts of Africa were thrown together as anonymous slaves to their own bewilderment and confusion, they desperately sought some compassionate and unifying spiritual force that could make a brutal existence endurable.

In one of history's cruelest ironies, they embraced their masters' religion, Christianity, as the only common creed available to them. And, in *all* cases, they embraced it with a fervor that generally amused whites (since whites did not share and may not be capable of sharing spiritual ecstasy, except for a few extraordinary individuals), while in *most* cases they seriously accepted the compassionate words attributed to Jesus at face value (which their masters did not) because they saw an apparent sameness between Christ's injunctions and the traditional social harmony they had known, and lost.

In their despair, they also equated the Jews' captivity in Egypt as paralleling their own enslavement in the New World.

Few of the African slaves could read or had any grasp of European historical reality. They could not be expected to know that the entire context of both the *Old Testament* and *New Testament* was really as anti-African as it is possible to imagine. Even after Emancipation in the United States, and even after some Blacks had achieved reasonable education, that African capacity for fervor and almost metabolic African need for social harmony and compassion blinded many Blacks to the real nature of Western Man's psychology *and blinded them to the fact that the various "white man's religions," deep down, simply reflected this violent and chauvinistic psyche.*

For, the fact of the matter is that the *Old Testament* and the *New Testament*—that *Holy Bible* upon which the Los Angeles jurors swore—is, *if it is considered objectively*, one of the most ethno-centric, anti-feminist, potentially racist and certainly violent documents ever penned by any group of humanity. *The Holy Bible* is, essentially, the saga of a uniquely wrathful and decidedly uncompassionate god who made a covenant with just one small tribe, the Hebrews, declaring them the universe's "Chosen People." You will look in vain for such arrogance, absurdity and *injustice* among the writings of Buddhism, Hinduism, Shinto, Taoism; you will look in vain for anything similar among the advanced African religions, of the Yoruba and Akan, that European slavery disconnected from Africans brought to the Americas. None of these non-white religions are nearly so racist, ethno-centric, anti-feminist and pathologically violent as the three Western religions (Judaism, Christianity and Islam) which all "devolved" from the *Old Testament* and the wrathful god of the Chosen People.

This truth is slowly reaching African Americans and even invading the ivied halls of institutions like Yale Divinity School. On April 24-25, 1992, I was invited to participate in a symposium called "Racism: A Crisis of the Heart" at Yale Divinity School in New Haven, Connecticut. At that symposium, I presented the gist of *The Iceman Inheritance* and *Chosen People From The Caucasus*: that

Western Man is basically more violent than other human groups because of significant Neanderthal ancestry; that racism, sexism, ethnocentrism and inordinate technological "progress" are all symptoms of this basic Neanderthal mindset; that the ancient Hebrews originated in a region of known Neanderthal concentration; the god and religion conceived by these early migrants into Palestine merely (naturally) reflected their own psychology. Christianity, as an offshoot of Judaism, like Islam, inherently retains much of the original ethnocentricity, anti-feminism and inhumanity ("believe in *me* or else") of that early Hebrew conception.

The focus of the Yale symposium was the fervent, but unfounded, assumption by African American clerics that Christianity was the only religion capable of defeating racism throughout the world; they held the symposium to develop ideas on strategy. But, in my view, which I tried to defend with actual evidence, not hopeful belief, was that Christianity is *and always was*, fundamentally racist, ethnocentric, anti-feminist and that the god of the *New Testament*, while not as violent as the god of the *Old Testament*, contained plenty of "believe in me or be damned" intolerance and *inhumanity*. Other major religions of the world, and traditional African ones, simply do not incorporate or embrace "exclusiveness/chauvinism" and "violence" to any comparable degree. *They* (pick any one), *not Christianity*, are better conceptual/spiritual orientations for combatting racism and could be embraced with fervor by African Americans to better effect.

This view was, of course, a "heresy" in the eyes of the traditionalist, fundamentalist African American Christians in the audience—and there were a number who were shocked. However, there were also a number of younger

"Christian" seminarians, some studying at Yale and some who had come to the symposium from other divinity schools, who were not so shocked. They had come to the same conclusion, but on spiritual and simple observational grounds. They welcomed a presentation of the biological and anthropological "proof."

The major, hurtful lesson for African Americans in the Rodney King affair is not simply that white society, and white police who protect it, are needlessly brutal because of Neanderthal ancestry—I think that most progressive Blacks know that they live in a "culture of violence" (as Penny Hess depicts Western Civilization in her book of the same title)—but an even worse realization: that even the spiritual, religious and judicial concepts of white society, all originally articulated in basic form in the *Old Testament*, are fundamentally "against" and unsuited to African American human rights and aspirations. Those jurors swore on the *Bible*, but it is a white covenant...and it always was.

Chosen People From The Caucasus presents data that merely prove, and possibly explain, the existence of a racial/cultural reality. It is not intended to suggest solutions posed by that reality. Nonetheless, the too-believable injustices of the Rodney King affair has angered me so much that I cannot help but make a plea: African Americans must forsake the white man's social structures, concepts of justice and, yes, even religion and return, as far as possible, to genuine African values and identity (insofar as these can be accurately recovered and reconstructed). I make this plea, not only for you who are Black, but for myself (and my son, and other people) *who are white.* Some of us realize that we are all living in a culture of violence, and we would prefer an alternative; some of us have fought, thought and written in the hope of attaining an alternative. We are "white," but non-whites. All non-whites, including Asian-Americans and Hispanic-Americans of mixed Spanish/Black/Amerindian origins must strive to create some cultural alternative to the Western World's culture of violence that came from the Caucasus so long ago.

There are 30-40 million Blacks in North America and a return to traditional African values by so many could not be thwarted, defeated or eradicated by white supremacy. An unassailable cultural alternative would be created, and many non-Blacks would probably find greater security and compassion for their own values than anyone might suspect. African Americans have more allies than are evident from the Rodney King injustice, and we continue to look to Blacks to create an alternative to our culture of violence.

Michael Bradley
Toronto, Canada
May 1, 1992

*All information and quotes concerning the Rodney King affair have been taken from the *Toronto Sun's* coverage (April 30, 1992, page 1 and 2 based on *Los Angeles Daily News* reports, plus Reuters and UPI wire services) and the *Toronto Star* (April 30, 1992, page 1 and A18 based on *Los Angeles Times* coverage and Reuter/AP wire services).

NOTES

INTRODUCTION

1. Michael Bradley, *The Iceman Inheritance* (Toronto: Dorset Publishing Inc., 1978), p. 1. This and all subsequent quotes from this book derive from the original Canadian edition specified here. The first U.S. reprint, by Warner Books of New York (1981), is now out of print; various pirated editions have different page numberings; the current U.S. edition by Kayode Books of New York includes a certain percentage of copies in which the pages are out of proper order.

2. Last of four known glacial periods of the Pleistocene (roughly, the last million years) according to the "European Sequence," and the only one to concern us here: the supposed emergence of modern man (*Homo sapiens sapiens)* from *Homo erectus* and *Homo sapiens neanderthalis* is a complicated process still not completely understood or agreed upon.

3. Indeed, the current Kayode edition adds to the total of typographical errors in the material typeset, for this recent U.S. edition; also about 20-30% of Kayode copies are faultily bound with some pages missing and others out of proper sequence.

4. A noted Hungarian psychotherapist who worked closely with Freud.

5. As early as 1872, James D. Dana, *Coral and Coral Islands* (New York: Dodd, Mead) and then James Croll three years later, *Climate and Time* (London: Daldy, Isbister, 1875) had suggested that the last Ice Age had not involved a uniform cooling of the Northern Hemisphere. Freud could have been unaware of these American and British authors, but not of the work of Amadeus W. Grabau *A Textbook of Geology* (New York: Heath, 1920) who originally worked out of Vienna and whose original German edition of the above work was published in 1918. Ellsworth Huntington (1907); William H. Hobbs (1905, 1911) and Carl Freiherr Loffelhoz von Colberg, *Die Drehungen der Erdkruste in geologischen Zeitraumen* (Munich, 2nd edition, 1895) were available to Freud, as well as reports on recovered Siberian mammoths by Frederick A. Lucas (1899)

and Richard Lydekker (1899), which emphasized that the Siberian climate had been warmer than today when the mammoths had lived during the last Ice Age. Since Freud was extremely interested in the "Ice Age," it is most unlikely he was ignorant of the fact that some regions of the Northern Hemisphere were warmer during it than today. It was a topic of puzzled interest from 1900-1920.

6. Frances Cress Welsing, *The Isis Papers: The Keys To The Colors* (Chicago: Third World Press, 1991).

7. Cheikh Anta Diop, *Civilization or Barbarism: An Authentic Anthropology* (Brooklyn: Lawrence Hill Books, 1990).

8. (New Brunswick, New Jersey: Rutgers University Press, 1987).

CHAPTER ONE: THE CHOSEN PEOPLE

1. Michael Bradley, *The Cronos Complex* (Toronto: Nelson, Foster & Scott, 1973).

2. *The Iceman Inheritance*, p. xviii, Posner's Foreword.

3. See Montagu's *Man's Most Dangerous Myth: The Fallacy of Race* (New York: Oxford University Press, 5th edition, 1974).

4. See Carleton Coon's *The Origin of Races*, (New York: Alfred A. Knopf, 1962) for his ideas about the original geographic domains of all five of his major races. He associates the Australoids with the Pithecanthropus hominids of ancient Indonesia, pp. 372-373, Map 9.

5. Ibid., p. 460-461, discusses the seventeen separate diagnostic characteristics found in Mongoloid skeletal material that differentiate this race from all others (quoting Franz Weidenreich and Davidson Black who did the initial physical anthropological assessment using the actual bones of Sinanthropus "Peking Man"). See also Harold L. Shapiro's *Peking Man* (New York: Simon and Schuster, 1974) pp. 72-73 where these characteristics are given a more brief discussion, Weidenreich quoted again, and current Chinese opinion given.

6. Montagu, *Man's Most Dangerous Myth: The Fallacy of Race*, p. 37.

7. I would amend this only to say that some Native American and Eskimo groups today speak virtually only English or French, and for them, in conflict with the major social authorities, the conflict *is* essentially racial; it is further clouded by the language issue.

8. See Ivan T. Sanderson's, *African Treasure* (New York: Viking,

1935) pp. 117-119 for a description of this oratory and a comment on the African love of litigation.

9. Michael Bradley, *The Columbus Conspiracy* (Toronto: Hounslow Press, 1991) p. 170 where these four principal backers are listed; they are also named in this book.

10. Ibid., pp. 212-213, quoting from Steiner's *Dark and Dashing Horsemen* (New York: Doubleday, 1974) but with some original references checked.

11. Ibid.

12. These figures quoted come from a personal conversation with Dr. Ray Winbush of Vanderbilt University, September 21, 1991, where he provided these figures from his extensive reading and research.

13. From Winbush, as above, I find it difficult to believe that 100 million could have persished, considering the entire world population during the Renaissance and the number of European oceanic ships in register.

14. Information given to me verbally by Dr. Molefi Asante on October 4, 1991 who has researched this matter; but see also the text of Dr. Leonard Jeffries, July 20, 1991, speech quoted elsewhere in this text and notes.

15. Felix Warburg headed the U.S. Federal Reserve System; Saul Warburg headed the German National Bank. See *Columbia Viking Desk Encyclopedia* entry for "Federal Reserve System" p. 356, and for "Warburg" p. 1152.

16. Edward Charnay, *The Birth of the Bomb* (London: Rupert-Davis, 1962) p. 186. This Einstein letter to Roosevelt is reproduced in full.

17. Canada has experienced at least two major "events" at Chalk River in which significant radioactive pollutants escaped into the Ottawa River; John G. Fuller recounted the near melt-down at the Enrico Fermi breeder reactor near Detroit: *We Almost Lost Detroit* (New York: Reader's Digest Press, 1972). But see also Fuller's *The Day We Bombed Utah* (New York: New American Library, 1984).

18. Edward Charney, *The Birth of the Bomb*, p. 138.

19. Ibid, pp. 19-21.

20. Ibid, pp. 138-141.

21. Ibid, p. 142, particularly the great Dnepropetrousk hydroelectic dam.

22. Ladislas Farago, *The Game of Foxes* (New York: D. McKay,

1971) pp. 64-65.

23. Neal Gabler, *An Empire of their Own: How the Jews Invented Hollywood* (New York: Doubleday, 1989).

24. Personal correspondence with Dr. Ray Winbush who informed me of this (from Nashville newspaper articles) on January 20, 1992.

25. Dr. Leonard Jeffries, Speech of July 20, 1991 as reprinted (entire text) in the New York *Newsday* pp. 3 and 25-29 (August 19, 1991). Hereinafter "Jeffries, *Newsday.*"

26. The *New York Times* (August 7, 1991), p. B1.

27. Jeffries, *Newsday*, Col. 2, p. 25.

28. *Emerge* magazine, "Melanin Madness" (February 1992), p. 32.

29. Personal communication from Dr. Ray Winbush, December 20, 1991.

30. Calculations based on Arthur Koestler's figures given for 1960 world Jewish population *The Thirteenth Tribe* (London: London & Co., 1976), p. 181.

31. Robert Ardrey, *The Territorial Imperative* (New York: Antheneum, 1966), pp. 56-62.

32. Konrad Z. Lorenz, *On Aggression* (London: Methuen & Co., 1967). Quoted here from Nethen's University Paperback edition, "The Pair Bond," pp. 141-148 (geese) and about jackdaws in *King Solomon's Ring* (New York, New Library American, 1972).

33. Ardrey, *The Territorial Imperative*, p. 110.

34. Bradley, *The Cronos Complex* (Toronto: Nelson, Foster & Scott, 1973), p. 32.

35. Dian Fossey, "Imperilled Giants of the Forest," *National Geographic*, Vol. 159, No. 4 (April, 1981), pp. 501-522.

36. C.R. Carpenter, "A Field Study of the Behavior and Social Relations of Howler Monkeys," *Comparative Psychological Monogram* (1934), pp. 1-168. About territorial assertion of howler monkeys in general, but mention of "Napoleon" and the Rhesus groups on pp. 76-81.

37. Skeat and Blagden's *Pagan Races of the Malay Peninsula*, (London, 1906), Vol. ii, pp. 281-283. One of the classic "white man's burden" books.

38. Eugene Marais, *My Friends the Baboons* (London: Methuen, 1939), pp. 159-178, particularly. In this work, actually one of the first field studies, Marais stated his belief baboons have a "phyletic memory," or some notion of their racial existence *in the past.*

39. Dian Fossey, "Imperilled Giants," *National Geographic*, cited previously, stated her initial belief that murder of some offspring was a measure taken to avoid inbreeding. But her belief changed with observation. Page 512 of this article, by Fossey "...I now believe that infanticide is the means by which a male instinctively seeks to perpetuate his own lineage by killing another male's progeny in order to breed with the victim's mother." Now, it may be that Dian Fossey did not fully grasp the real psychological urge here. If a male wished merely to perpetuate his lineage, yes, he'd murder the offspring of a displaced male rival—but he could then perpetuate his lineage by breeding with any and all other available females. To breed purposefully with the victim's mother indicates something beyond the mere wish to perpetuate his own lineage; it indicates a compulsion to *erase* his displaced rival's identity as thoroughly as possible: the rivial and offspring are dead, and his offspring will issue from the same female. It is as though the rival and offsping never existed. They have been fully and completely substituted with his own lineage, through the same female. To me, this indicates some dim awareness of future identity—that is, a glimpse of "identity-in-time" that is characteristic, or even "definitive" of fully human culture.

If this idea is at all defensible, then one can take issue with a basic tenet of Sigmund Freud's "psychosexual theory of personality development." The basic human motivational drive in terms of personality development must be to acquire and preserve a secure identity *in the dimension of time.* True, the means of doing this may be highly sexual in content, especially with infants and young children who have no other means of communication, but the basic urge is, essentially, "identity-in-time," not sexual assertion. It seems to me that a "psychotherapy" based on this premise would not only reflect observed behavior of apes and men more accurately, but might get better results.

40. Ardrey, *The Territorial Imperative*, pp. 106-107. But see also J.S. Brown's *Motivation of Behavior* (New York: McGraw-Hill, 1961) for similar experiments.

41. Ardrey, *The Territorial Imperative*, p. 108 and p. 170.

42. Ibid, p. 106.

43. Bradley, *The Cronos Complex*, p. 40; but see also chapter "Homo Temporalis" p. 42-63.

44. Bradley, *The Iceman Inheritance*, p. 111.

45. Ibid., p. 110-125.

46. Bradley, *The Cronos Complex*, p. 54.

47. Ibid., p. 42-63.

48. Alexander Marshack, *The Roots of Civilization* (New York, McGraw-Hill, 1971).

49. Bradley, *The Iceman Inheritance*, p. 110-125.

50. Desmond Morris, *The Naked Ape* (New York: Bantam, 1969).

51. This breast evolution is the subject of a psychosexual phantasmagoria of mine, perhaps mercifully unpublished, entitled *Kephra* (and later) *Persephone's Hem.*

52. Morris writes at length about this in *The Naked Ape*, pp. 128-163, but see also Shapiro's humorous conclusions on female buttocks, *Peking Man*, pp. 140-141.

53. See Coon's *The Origin of Races*, p. 639, quoting from Biasutti's original discovery of Capoid racial traits in Egyptian mummies.

54. Laurens van de Post, *The Lost World of the Kalahari* (Harmondsworth: Penquin, 1958), p. 32.

55. One has only to read the various versions of the *Kama Sutra* to appreciate the Oriental appreciation of tactile stimulation and, although the *Kama Sutra* variants were actually written in India, it is generally agreed that it represents an imported Chinese book now lost.

56. This bit of information (myth?) comes not from anthropological surveys and measurements in the field, but from Dr. J.B. Webster of Dalhousie University's Department of African Studies, a man who spent many years in Africa doing ethnographic research. Personal communication, April 1979. Being a student of Dr. Webster was, well, interesting.

57. Bradley, *The Iceman Inheritance*, p. 123 for a brief discussion of these cave paintings.

58. Which is to say that only among humans does this powerful urge to attain identity within the "territory of time" reach a full recognition and expression, although—as Dian Fossey observed with her murderous gorrilla—the highest primates seems to have a vague notion of *future* identity in time; Eugene Marais maintained that baboons may have a grasp of some *past* existence of their kind in time. But only among humans is there a complete "three tense continuum" of existence comprising past, present and future.

CHAPTER TWO: NEANDERTHAL-CAUCASOIDS

1. "Family Tree: mother of modern humanity traced to Africa, lived 200,000 years ago," Associated Press, Washington (September 27, 1991), the announcement of Dr. Marc Stoneking, an anthropologist at Pennsylvania State University. The study by Stoneking's team involved mitochrondrial DNA (mtDNA, for short) which is DNA found outside the nucleus of cells and is found only in females. The technique of "DNA distancing," or determining evolutionary relationships from mutations in DNA, is based on several assumptions. First, that the DNA within an entire species is, at some point that can be ascertained or inferred, identical in structure within random individuals. Second, that mutations within this cohesive and originally identical DNA structure also take place at random intervals of time. Therefore, in theory, from any assumed starting point of DNA composition, researchers can compute, statistically, the time required to allow the DNA to "evolve" a given number of random mutations. The technique of using mtDNA was invented by Dr. Rebecca Cann and Allan Wilson of the University of California, Berkeley, in the late 1970s and early 1980s. From this sort of analytical method, Stoneking and others compute that "modern humanity" evolved from something, and did it in Africa "somewhere south of the Sahara Desert" about 166,000 to 249,000 years ago.

However, from this methodology, which they invented, Cann and Wilson had earlier (1987) released somewhat different findings: that "man" only differentiated from the African anthropoids (chimpanzee and gorilla, but closer to the chimpanzee) about 5 million years ago; and that three primary races (Mongoloids, Negroes and Caucasians) differentiated about 300,000 years ago. See *The Monkey Puzzle* (New York: McGraw-Hill, 1982).

Although enthusiastic Afrocentrists have grabbed onto all of this with glee, and usually (at least, judging from the ones I discussed this with) without an inkling of the problems involved with the DNA distancing method, the situation is far from cut and dried. First, and maybe least important, is the definition of "modern man" that people like Stoneking have used blithely without clarifying. Since classical physical anthropology shows, without too much room for doubt, that morphologically "modern people" were alive and well and living in Europe and elsewhere earlier than Stoneking's computations indicate, then what are we to make of that? Then, according to earlier mtDNA distancing studies made by the inventors of the method, human races had differentiated 100,000 years *before* Stoneking's first (hypothetical) "modern woman." What race was she? And, were the other races "modern"?

Further, it is questionable that anyone, with any claim to accuracy at all, can pick a "point in time" where the DNA would have been stable, consistent and identical among ancestral human groups. Added to that is the *assumption* that this DNA would, statistically, mutate randomly—which is to say, from the statisticians's point of view, at a more or less constant rate that can be timed. Evidence is building that both geological and biological evolution takes place in sudden, "catastrophic" changes at unequal intervals. The "Doctrine of Uniformitarianism" is crumbling—the whole point of Chapter 3 of *Chosen People From the Cacasus*. Therefore, a species' DNA could have remained virtually unchanged for hundreds of thousands of years, maybe even millions of years, and then suddenly mutate within just one or two generations *because of heat or other radiation* to produce a new organism type, but one obviously related to a previous gene pool.

At this point, the claims made for DNA distancing do not conform to evidence from other sources; and the method may be founded upon unwarranted assumptions. One of the top physical anthropologists in the world, Dr. Eric Trinkhaus of the University of Mexico, calls the Stoneking study a "house of cards," a collection of unverifiable and perhaps unwarranted statistical assumptions. I tend to agree, from the experts I have consulted, that the technique of DNA distancing will probably prove to be very valuable—once the bugs are worked out—but that the present spate of headline-making announcements about the evolution of man (or, rather, of "woman") will turn out to be embarrassing errors made by enthusiastic pioneers eager to promote their science.

Unfotunately, the entire "race" and "human evolution" issue is so political that people with no technical knowledge of DNA distancing will simply accept Stoneking-type sensationalism because of these flatter prejudices: Afrocentrists will be flattered, feminists will be flattered—and so we'll have to live with this distortion for some time.

I have no objections to the probability that more or less modern man (if we can define the term) may well have evolved in Africa; and no objections to the 50/50 chance that this first specimen thereof may be female but I suspect that she/he will be considerably older than 250,000 years.

2. *The Monkey Puzzle,* pp. 252-255, note research of Becky Cann.

3. This is precisely what C.S. Coon argues in *The Origin of Races*, p.460-461. His opinion is backed by those of Ranz Weidenreich, Harold Shapiro and by almost all Chinese anthropologists who have worked with Sinanthropus ("Peking Man") remains.

4. See Coon's *The Origin of Races,* p. 508, quoting B. Skrij and other anthropologists.

5. Coon wants to make them "Caucasoids," but on what evidence? *The Origin of Races*, p. 511.

6. Charles Hapgood, *The Path of the Pole* (Philadelphia: Chilton, 1970), pp. 93-102.

7. Bradley, *The Iceman Inheritance*, pp. 70-74.

8. In *The Origin of Races*, Coon explains "Allen's Rule" in great detail, p. 60.

9. Ibid., p. 529.

10. Ibid., p. 548.

11. Ibid., p. 529.

12. Ibid., p. 533.

13. Ibid., p. 547.

14. Bradley, *The Iceman Inheritance*, pp. 87-125.

15. Ibid., pp. 95-102.

16. Ralph Solecki, *Shanidar, The First Flower People* (New York: Alfred Knopf, 1971); but also see Solecki's "Three Adult Neanderthal Skeletons from Shanidar Cave in Northern Iraq," *Smithsonian Report,* Pub. No. 4414), pp. 603-635.

17. Coon, *The Origin of Races*, p. 545.

18. Ibid., p. 639.

19. Ibid., p. 520.

20. Ibid., pp. 500-533

21. Ibid., p. 335, chart of Pleistocene chronology and fossil humans.

22. C. Emiliani, *Science* (Vol. 123, 1956), p. 925.

23. I have in mind here H.G. Wells, *Outline of History* (London, 1920). He stated on p. 47 that Cro-Magnons would have disdained breeding with Neanderthals because the stench of Neanderthal caves would have offended Cro-Magnon's evolved sensitivities. I doubt it.

24. *The Toronto Sun* (June 27, 1991), p. 63. Very advanced modern people have been enthusiastic cannibals, such as the Polynesians. No need

to attribute cannibalism to the primitive nature of Neanderthals. All ancient peoples were probably cannibals to some degree, but for ritualistic rather than dietary purposes.

25. *Columbia Viking Desk Encyclopedia*, entry for "Basque Provinces," p. 91.

26. Coon, *The Origin of Races,* page 547.

27. Ibid., p. 545.

28. Indeed some of these plaster Neanderthal reconstructions bear a surprising individual resemblance to actual people I have met who are of Russian, European Jewish or Transcaucasus extraction. I remember once sitting in a restaurant in St. Catherines, Ontario—a town that has many Slavic immigrants—when my attention was drawn to a most remarkable couple. The man was barely 5'5" and his wife noticeably shorter; they were both stout and muscular and had large faces with very large noses, sloping foreheads and rather pronounced brow ridges. They looked, for all the world, like illustrations of gnomes in a children's book. Doing some "mental measuring" based on my training in physical anthropology, it was obvious that, in some respects, this couple's physical traits were more extremely Neanderthal than many skeletons in museums. I engaged them in conversation (they were very pleasant), and learned that they both came from Tiflis, in the Georgian Caucasus, but that their families were mountain folk from a village I have never been able to find in any atlas. They were pleasantly well aware of the reason for my curiosity (everyone else studiously avoided looking directly at them, and snatched surreptitious glances) and told me that many people like themselves lived in the mountains, although they kept to themselves; they also told me about the *kaptar* of the Caucasus. They operated a profitable nursery in St. Catherines (a town noted for its flower industry), and we eventually parted outside on the sidewalk. They seemed to be a cheerful and jolly couple, but lonely.

29. Coon, *The Origin of Races,* pp. 582-585.

30. This is a mythical distortion. Neanderthal man did not have a well-developed chin. See Coon's *The Origin of Races,* p. 536, and the profiles of the Neanderthal skulls reproduced in *The Iceman Inheritance* (after Coon) p. 80.

31. Ivan T. Sanderson, *Abominable Snowmen: Legend Come to Life, the Story of Sub-Humans on Five Continents from the Early Ice Age Until Today* (Radnor: Chilton, 1951), pp. 284-287. Khakhlov's story on pp. 307-

317.

32. I confess that I was a bit nonplussed when I read this last bit, having always thought of tigers and leopards as being tropical beasts of India, Southeast Asia and Indonesia, but it turns out that these cats originated in Central Asia (where the largest tigers are still to be found—the Ussuri tigers of Siberia) and migrated toward Eurasia very early on, and only much more recently down into India, Bengal and the tropics. Bernard Heuvelmans, *On The Track of Unknown Animals* (London: Rupert Hart-Davis, 1958), p. 335.

33. Sanderson, *Abominable Snowmen*, p. 276.

34. James A. Michener, *Caravans*, well researched novel and movie.

35. Heuvelmans, *Unknown Animals*, p. 28.

36. Sanderson, *Abominable Snowmen*, p. 319.

37. Ibid., p. 11.

38. Ibid., p. 488.

39. Solecki, *Shanidar*, pp. 1-6.

40. Sanderson, *Abominable Snowmen*, p. 334.

41. Ibid., pp. 295-296.

42. U. Zerchaninov, "Following the Footprints of Kaptar," *Komsomol Youth* (August 29, 1959), pp. 21-27. I find it unlikely that the Neanderthals had no language.

43. Sanderson, *Abominable Snowmen*, p. 380, footnote to statement by Rabbi Yonah N. ibn Ahron. This statement by Ahron is interesting because this description of scent cannot have survived from biblical times. The potato was unknown in the Old World before 1500 A.D. Since then, it has been grown widely in Eurasia, Central Asia and North America—a "smell like a field of potatoes" is often attributed to primitive hominids in the region. But, and this is an important point, if the exegetic literature applied this potato smell to Esau of tradition, then it must have been a late embellishment to the Esau story, one developing after, say, 1600 A.D. when potatoes had reached Eastern Europe. This same scent description has been applied to the kaptars—so, in describing Esau the same way, the Jewish historians were obviously, in "hindsight," compar-

CHAPTER THREE: THE ICEMAN INCARNATE

1. Hapgood, *The Path of the Pole*, pp. 288-289.

2. *Columbia Viking Desk Encyclopedia*, entry for "James Hutton," p.495.

3. His name was Walter Bucher. Hapgood, *The Path of the Pole*, p.197.

4. Hapgood, *The Path of the Pole*, page 160 where Emiliani's chart of "Temperature Variation of the glacial Pleistocene" (1956) is reproduced starting with the cold peak of the Gunz glaciation circa 300,000 years ago to the present. There may have been another Pleistocene glaciation before Gunz in the European sequence; no one is certain. It is interesting to note, in this context, that Coon dates the Mauer mandible (ie. "Heidelberg Man") at circa 450,000 B.C., or 150,000 years before the glacial peak of the first European glaciation. Heidelberg Man lived during a very warm European climate, and Coon classifies him as *Homo erectus*.

5. Hapgood, *The Path of the Pole*, pp. 60-88, Chapter 3. "Failure to Explain Climatic Change."

6. Ibid., pp. 60-68.

7. Ibid.

8. Ibid., pp. 63-67.

9. Ibid.

10. Ibid. See also Heuvelmans, *Unknown Animals*, pp. 340-347 for a warmer Siberia in relation to mammoth stomach contents.

11. Ting Ying H. Ma, *Research on Past Climate and Continental Drift Based on the Growth Rate of Corals (Ordovician)*, Vol. I. privately published (Fukien, 1943) Vol. II (Silurian), privately published (Fukien, 1943); Vol. III (Devion) (Fukien, 1943); Vol. IV "The Equator and the Relative Positions of the Continents During the Cretaceous period as Deduced from the Distribution and Growth Values of Reef Corals" (Taipei: Taiwan National University, 1951); Vol. V (Since the end of the Cretaceous, (Taipei: Taiwan National University, 1952); Vol. VI "The Sudden Total Displacement of the Outer Shell of the Earth Relative to the Fixed Rotating Core the the Earth" (Taipei: Taiwan National University, 1956).

In over 70 maps, plus diagrams, and illustrations, Ma traces the past coral-growing zones from the Ordovician Period (circa 420 million years ago) to the present. Since corals can grow only in a narrow tropical zone, Ma was then able to supply the past locations of the North and South poles and the past locations of land masses from which no coral specimens could be obtained.

This massive, but no less meticulous, body of work is available in its entirety in North America only at Columbia University. I have consulted these volumes (1967-1969), but here rely on Hapgood's assessments/illustration as reproduced in *The Path of the Pole*.

12. Hapgood, *The Path of the Pole*, pp. 1-3, quoting James Clerk Maxwell and Sir George Darwin.

13. According to the "Tidal Theory" of solar system formation advanced by James Jeans and Harold Jeffries (1918 and later). *Columbia Viking Desk Encyclopedia*, p. 1087.

14. Personal correspondence with Dr. Dale Russell, National Museum of Man, Ottawa, Canada, June 1987.

15. Hapgood, *The Path of the Pole*, pp. 68-71.

16. Hapgood's *The Path of the Pole* was basically a revised edition of *The Earth's Shifting Crust* (1958) jointly authored by Hapgood and James H. Campbell. However, Albert Einstein's original foreword and endorsement are reprinted on pp. XIV-XV of the 1970 Chilton edition of Hapgood's *The Path of the Pole*.

17. Hapgood, *The Path of the Pole*, p. 3.

18. J. Tuzo Wilson, "Continetal Drift" (Reprint), *Scientific American* (April 1953).

19. In this theory, the area transported into a new polar zone would not suffer much glacial scouring at the actual center of the newly-forming ice cap. Scouring would only occur when the newly-formed ice cap reached sufficient mass to begin flowing outwards. This is, in fact, precisely what was discovered about the last (Wisconsin) ice sheet in North America—see Hapgood, *The Path of the Pole*, p. 130, quoting A.P. Coleman about the lack of scouring around Hudson Bay (1929) during the Wisconsin glaciation.

20. All of the great oceanic currents are circular in formation. They flow clockwise in the Northern Hemispehere, and counter-clockwise in the Southern Hemisphere. They are, in fact, giant "heat transfer" agents. They transfer equatorial heat to the arctic (like the Gulf Stream in the Atlantic, and the Kuroshiwo in the Pacific) and then take arctic water to the equator (where it is gradually warmed), like the Humbold Current of the south Pacific that allows penguins and sea lions to live on the tropical Galapagos Islands. Each polar ice cap is affected by these currents. But, obviously, a polar ice cap on water will be much more affected by heat transfer than an ice cap essentially, or mostly, on land. The reason why

the arctic ice cap is so much smaller than the antarctic one is simply that the North Pole presently happens to be located in an ocean, where warming currents can directly affect ice formation.

21. But the antarctic ice cap is located above a group of a large islands which insulates it from the direct influence of warming oceanic currents. It is therefore much larger than the north polar ice cap.

22. Hapgood, *The Path of the Pole*, p. 331 and illustration. The center of mass of the antarctic ice cap is actually 345 miles from the pole.

23. Hapgood, *The Path of the Pole*, pp. 330-337. The tangential force generated is some 6.82766×10^{12} metric tons, according to James II. Campbell's calculations, which were verified by Albert Einstein.

24. Hapgood, *The Path of the Pole*, p. 24.

25. Although the continents drift much too slowly to explain the Pleistocene Ice Ages in Europe and America, it is true that about 90 million years ago, Europe and North America were close together within the same great land mass because they shared the same species of dinosaurs. However, by about 80 million years ago there was apparently a widening gap between the two land masses that had grown sufficiently to prevent the Old World's iguanodons from reaching North America, and at the same time prevented the North American ceratopsians from reaching Europe.

26. Hapgood, *The Path of the Pole*, pp. 337-351 (including notes pertaining to a conference with Einstein).

27. Hapgood, *The Path of the Pole*, pp. 81-87 where Ma's Ordovician map of coral formations is reproduced (p. 82), Ma's Silurian map (p. 83), Ma's Devonian map (p. 84), Ma's Lower Carboniferous map (p. 85) and Ma's Late Cretaceous map (p. 86). These early maps show one pole in mid-Pacific and the other in or near present South Africa. For all maps, Ma has supplied an equatorial line *but has not attempted to show continental drift.* When this is taken into account, both poles would have been located over oceanic regions for considerable periods of time during the Paleozoic years ago. During this immensely long period of about 240 million years, the earth's climate may have been milder and more stable. When gradual shifting of polar regions began to bring them over land about 180 million years ago, climatic fluctuations began to be more severe. Since the Cretaceous, about 72 million years ago, the poles have often included areas of alternating land and water which has caused apparently frequent disruptions; but above all, the past million years have seen an acute

situation of polar regions over various combinations of land and water. This has caused (according to Campbell's theory) frequent and acute crustal shifts that have changed the climate of certain regions dramatically and suddenly about every 50,000 to 70,000 years.

28. Ibid., pp. 81-87, explanation as above.

29. Ibid., explanation as above.

30. Ibid., pp. 94-95, a chart showing difference between temperatures in the European Sequence and the North American sequence of the "Wurm-Wisconsin" glaciation.

31. Ibid., pp. 87-124.

32. Ibid., pp. 36-39 and Fig. 10 on p. 37.

33. Peter V. Glob, *Al-Bahrain* (Copenhagen, 1968), pp. 214. See also Gooffrey Bibby's *Looking for Dilmun* (New York, 1969), p. 253.

34. Bibby, *Looking for Dilmun*, pp. 191-205.

35. Ibid., pp. 253-257.

36. Paul Hormann, *Conquest by Man* (New York: Harper, 1954), pp. 112-124.

37. Eberhard Stechow, "Wann kamen die Malien zucrst nach Madagaskar?" *Forschungen und Fortischritte* (Berlin, No. 18, 1944).

38. Hapgood, *The Path of the Pole*, pp. 87-124.

39. Altitude is as much an agent of local climate as latitude, which is why there can be snow-capped mountains near the equator. For a lucid discussion of this, see Sanderson's *Abominable Snowmen*, pp. 398-420 and climate zone map, p. 419.

40. None of the Old World ungulates (sheep, goats, cows, antelope, etc.) occurred in the South American fauna and so would not have been available for domestication by man.

41. Taiwan once had some of the Old World ungulates, but not the horse nor several plants: cotton, coffee, date palms.

42. Coon, *The Origin of Races*, p. 635.

43. Ibid., p. 485.

44. Ibid., p. 18, Australoids in central India indicates past wide distribution.

45. Ibid., pp. 649-651.

46. As of about 2000 years ago, the Malay Peninsula and Indonesia was the location of a number of glittering and brilliant maritime cultures and civilizations. These people traded from India to China in a variety of ocean going craft, see *Herman's Conquest by Man*, pp. 154-157. If these

people had arrived in Madagascar around the start of the Christian Era, they would have contributed their formidable maritime skills to the just-developing Arab-Zanj maritime culture of that very region. They did not, so they must have arrived in Madagascar before they developed sophisticated ocean ships.

47. Thor Heyerdahl, the *Tigris Expedition* (New York: Doubleday, 1980), pp. 1-28 (of the New American Library "Plume" paperback edition) gives an excellent description of the Marsh Arab lifestyle with color photos.

48. Thor Heyerdahl, *The Ra Expedition* (New York: Doubleday, 1971). *Ra I* was made by Baduma tribesmen of Lake Chad along the lines of their traditional *kadays*, used today for transporting cargoes and families on the lake. This first vessel began to break up and was abandoned in the western Atlantic. *Ra II*, slightly smaller, was made of African papyrus reed, but built by Aymara Indians from Lake Titicaca whom Heyerdahl flew to Morocco. This second reed shop proved to be extremely seaworthy and completed the crossing from Morocco to the West Indies without incident.

49. Heyerdahl, *The Tigris Expedition*, pp. 22-26, quotes from the Utnapishtim Deluge Epic of Sumeria about reed "arks".

50. These reed rafts, called gares by Marsh Arabs, are limited in size only by the availability of reed. Even today, some are quite large, not to carry cargo, but to be cargo. Reed rafts are made into gares which are then floated down to paper mills on the Tigris; they are taken apart and their reeds used for making paper pulp.

51. See *The Holy Bible, Authorized King James Version* (London: Harwin Press) and the *Bible Dictionary and Appendices*, pp. 1-79, appended to the end of the Biblical text. See entries for "gopher" and "ghittim" in this dictionary. The *Bible Dictionary and Appendices* of the Harwin edition are useful and accurate, having been compiled by a committee of noted Biblical experts; unfortunately, however, the page numbering is non-sequential for the *Bible Dictionary* and the following Appendices whose pages are not numbered at all! I will refer to this as *Harwin Bible Dictionary* in future citations.

52. Papyrus is now extinct in Egypt and does not occur naturally. The Papyrus Institute maintains some reeds in ponds in order to produce paper. It is thought that either some bacterial infection, or industrial pollution, has killed off Egyptian papyrus. See *The Ra Expedition*, pp. 120-126.

53. Heuvelmans, *Unknown Animals*, pp. 331-353, for several accounts of edible mammoth meat from corpses that had only recently began emerging from ice or permafrost. This proves that long ago, the mammoths were quick-frozen in a matter of hours and never once thawed until their modern discovery or their meat would have spoiled. Since it is now known that mammoths were not arctic animals, but lived typically in the subarctic tundra and dwarf forest, the conclusion must be that a crustal shift carried many thousands of them in a northward direction. They were transported from their tundra grazing grounds into freezing, arctic conditions, in a matter of hours. Their flesh had no time to decompose, and even their stomach contents of tundra grasses and meadow flowers are sometimes found in permafrost above the arctic circle; and the nearest tundra grasses and dwarf willow forests are at least 500 miles to the south. This indicates that the last segment of the crustal shift, perhaps 1000-500 miles in length took only a few hours to complete.

54. Hudson's Bay is surrounded on the east, south and west by the Canadian provinces of Quebec oand Ontario, Manitoba and the Northwest Territories. To the north, the bay is almost enclosed by Southampton Island and huge Baffin Island. Because of its nearly enclosed situation, Hudson's Bay is minimally affected by oceanic currents. An arctic ice cap centered on it would, therefore, likely be larger and colder than the present arctic zone.

55. One has to manipulate a globe to appreciate this.

56. Hapgood, *The Path of the Pole*, pp. 227-233. The concensus of expert opinion, quoted in these pages, is that 10,000-12,000 years ago the average sea level was from 500-300 feet lower than today. However, in some places the land was also raised or lowered by the cataclysm of the crustal shift and so the relative sea level in any given place is a difficult matter to ascertain.

57. Bibby, *Looking for Dilmun*, pp. 234-241. The present island of Bahrain cannot have been the whole of Dilmun; indeed, it may have been only a region set aside for burials. The actual civilization of Dilmun must have been concentrated elsewhere, but nearby.

58. Thor Heyerdahl, *The Tigris Expedition*, p. 25, quoting Sumerian myth.

59. Henri Lhote, *A la Decouverte des Fresques du Tassili* (Paris, 1958).

60. This was the co-called "Golden Age" of the Greek poet, Hesiod;

the "Saturnian Age," and so on, of fond memory (see *Hamlet's Mill* by Georgio de Santilliana and Hertha von Detchend for many references from many cultures).

CHAPTER FOUR: IMAGES OF A GLACIAL GOD

1. Thor Heyerdahl, *The Tigris Expedition*, pp. 22-29.

2. Von Daniken popularized the "cosmic astronaut" idea, but it was first stated by two prominent scientists, one American and one Soviet: see *Intelligent Life in the Universe* by Drs. Carl Sagan and I. S. Schlovskii (San Francisco: Holden-Day, 1966).

3. This segment of Rabbi Ahron's long quote, and all other segments, comes from Ivan Sanderson's *Abominable Snowmen*, pp. 377-382, where Ibn Aharon's observations are published in full.

4. Sanderson, *Abominable Snowmen*, p. 194.

5. Leonard C. Woolley, *The Sumerians* (New York, 1965), pp. 7-8, 192-194

6. Harold Shapiro, *Peking Man*, pp. 64-67.

7. Principally along the Iberian Corridor from North Africa and on into Western Europe, archeology in Spain and France indicates a very large number of post-glacial immigrants. A lesser, but still significant, number of post-glacial immigrants from Africa entered Europe via a Tunisia-Sicily-Italy route, but seem to have stayed near the coasts of the Italian peninsula.

8. *Harwin Bible Dictionary*, p. 8. "Babylonians."

9. Woolley, *The Sumerians*, pp. 8-24.

10. Heyerdahl, *The Tigris Expedition*, p. 27 (footnote).

11. Ibid., p. 25.

12. Ibid., pp. 22-27.

13. *Harwin Bible Dictionary*, p. 66, "Semites."

14. Ibid.

15. Robert Graves, *The White Goddess* (London: Faber and Faber, 1961), page 236.

16. *American Heritage Dictionary of Indo-European Roots*, p. 807.

17. Ibid.

18. Ibid. States that "Indo-European...probably spoken during the neolithic period, but the culture in which it was spoken has not yet been archeologically identified. "The first farming evidence from this Caucasus-

steppe region dates to about 7,000 B.C., thus dating the neolithic in this area about then. However, there's little doubt that the neolithic actually began several thousand years before our present evidence indicates. A date of 10,000 to 12,000 B.C. for spoken Indo-European in its primal form is my own, can be disputed, but is probably not so far off.

19. Graves, *The White Goddess*, pp. 30-40, gives several translations.

20. Ibid., p. 201.

21. Ibid., p. 261.

22. *Harwin Bible Dictionary*, p. 5, "apple."

23. Graves, *The White Goddess*, p. 263.

24. Sanderson, *Abominable Snowman*, p. 380, but also note #42 for Chapter 2.

25. I have often wondered whether this "beast within" is an unconscious recognition of the psychosexually maladapted personality compelled by the identity-urge to assert and defend territory in time. This "guilt" may be somewhat more pronounced in Jews, judging from Jewish jokes and preoccupations with mother-son relationships, but it seems also to be reflected in the Christian concept of "original sin" as well as in many other human cultures (in diluted form). As I pointed out in *The Iceman Inheritance*, and before that in *The Cronos Complex*, the essentially human recognition of time as a kind of "territory" that can yield identity puts us in ego-conflict with our "past" (parental generation and ancestors) and also in conflict with our "future" (personal offspring and posterity in general) because the territory must be asserted and defended against past and future if it is to yield identity. From a purely biological perspective, *if these theories are at all valid*, every human is born in conflict dictated by "the cronos complex." Is this time-factored conflict against the rest of humanity the source of "guilt" and "original sin," etc., and might it not be stronger in highly Neanderthaloid peoples?

26. *Harwin Bible Dictionary*, p. 53, "Paul."

27. I know of several instances in which precisely this has happened; one case presently involves a personal acquaintance of mine. I plan to write her story for North American religious publications.

28. The subject of the Cathars, their Holy Grail and beliefs and very recent revitalization in Europe is the point at which the "racist" writings (for which I am known in the U.S.) "interfaces" with the Holy Grail and religious-history writings for which I am principally known in Canada.

Cathars were a heresy involving a melding of Islamic, Christian and Jewish thought and dogma; it was a religion that also avoided the anti-feminism of the three orthodox faiths. Their secret treasure was supposedly the Holy Grail. In *Holy Grail Across The Atlantic* (Toronto: Hounslow Press, 1988), I present evidence that the impetus to "discover America" was actually a search by these 13th century religious dissidents for a refuge from the Inquisition. In *The Columbus Conspiracy* (Toronto: Hounslow Press, 1991) I suggest that Columbus was part of the same religious "Grail complex" and that his task was to find a haven in a more southerly latitude; the initial refuge had been established circa 1400 in Nova Scotia.

29. Bradley, *Dawn Voyage: The Black African Discovery of America* (Toronto: Summerhill Press, 1987), pp. 31-32, where the account of Malian Islam by Ibn Batuta is quoted in brief.

CHAPTER FIVE: THE KHAZAR CONVERSION

1. I *Samuel* 16.
2. *Harwin Bible Dictionary*, p. 39, "Jebusites."
3. I *Samuel* 16-20.
4. I *Samuel* 25.
5. II *Samuel* 11. See also I Kings 15:5.
6. *Genesis* 27:30-35.
7. *Nehemiah* 13:26.
8. *Harwin Bible Dictionary*, see "Jerboam" p. 40, "Rehoboam" p. 62.
9. Ibid., pp. 42-43, "Kings of Judah and Israel".
10. Ibid., "The Hebrew Old Testament," Apendix.
11. *The Gospel According to Thomas* (New York: Harper, 1959) (ed. E.J. Brill), pp. v-vi.
12. Hermann, *Conquest by Man*, pp. 26-32
13. Bradley, *Holy Grail Across The Atlantic* (Toronto: Hounslow Press, 1988), pp. 178-179. See also Michael Baigent, Richard Leigh and Henry Lincoln, *The Holy Blood and the Holy Grail* (London: Jonathan Cape, 1982), pp. 245-253; also J.M. Wallace-Hadrill's *The Long-Haired Kings* (London: 1962), p. 238.
14. Hermann, *Conquest by Man*, pp. 42-43.
15. Baigent, Leigh and Lincoln, *The Holy Blood and the Holy Grail*,

pp.245-253.

16. *Harwin Bible Dictionary*, p. 41, "Judah."

17. Joseph Jacobs, *Jewish Contributions to Civilization, An Estimate* (Philadelphia: The Conat Press, 1920), The Jewish Publication Society of America, Chapter 6, "Jews and Commerce," pp, 191-196. Jacobs describes small Jewish trading groups that existed from the Mediterranean to China from the Christian Era to the late Middle Ages.

18. Baigent, et al., *The Holy Blood and the Holy Grail*, pp. 245-253.

19. Arthur Koestler, *The Thirteenth Tribe* (London: Hutchinson & Co., 1976), pp. 159-168.

20. Ibid., pp. 163-168.

21. Ibid., pp. 166-168, quoting Mieses.

22. Ibid., p. 174.

23. Ibid.

24. Ibid., p. 167.

25. Ibid., p. 173, quoting Mieses.

26. "Ashkenaz" is first mentioned in Genesis 10:3 as a son of Gomer. Gomer, in turn, was a son of Noah's third son, Japheth (*Genesis* 10: 1-3). Ashkenaz, therefore, does not rate within the traditional Hebrew lineage that the line of Noah's son, Shem, led to Abraham and the "Chosen people" by virture of Abraham's supposed covenant with God. "Ashkenaz" was supposed to be the ancestor of these people, only vaguely known in the Mesopotamian-Palestinian world of Abraham, who lived beyond the mountains (Caucasus-Elburz chain) to the far north.

27. Koestler, *The Thirteenth Tribe*, pp. 241-246 (Selected Bibliography") lists scores of Jewish and non-Jewish historians who had drawn attention to the Khazar origin of Eastern European Jewry before Poliak and Artamonov.

28. A staff librarian of the Public Libraries of Metropolitan Toronto informed me on February 3, 1992 that of 18 copies of *The Thirteenth Tribe* purchased in September 1976 to supply the entire system, all but two have disappeared.

29. Here I have used "Byzantium" as a synonym for Constantinople" which is not, strictly speaking, historically accurate, but will do for the purposes of this book without undue distortion.

30. Koestler, *The Thirteenth Tribe*, p. 60.

31. Ibid.

32. Ibid., p. 61, quoting from A. Sharft's Byzantine Jewry-*From*

Justinian to the Fourth Crusade (London: 1971), p. 88. Strictly speaking, this description refers to Jewish persecution in the Italian domains of the Byzantine Empire (particularly under Basil I), and not to methods of torture used in Byzantium itself at roughtly the same period.

33. Ibid., p. 60.

34. Ibid., pp. 63-64.

35. Ibid., from the non-numbered title page, but taken from Mugaddasi's *Descriptio Imperio Moslemici* ("Description of the Moslem Empire") of the 10th century.

36. Identified by various medieval chroniclers; excavated by Artamonov.

37. Koestler, *The Thirteenth Tribe*, p. 62.

38. Ibid.

39. Ibid., p. 29.

40. Ibid., p. 26, quoting Gibbon (*The Rise and Fall of the Roman Empire*) based on Theophanes.

41. Ibid.

42. Ibid.

43. Ibid., p. 173, quoting Mieses.

44. Ibid.

45. Ibid., p. 176.

46. Ibid., frontispiece map.

47. Ibid., p. 18, footnote based on W.E.O. Allen's *A History of the Georgian People* (London, 1952).

48. Ibid., p. 157.

49. See the *Shorter Oxford English Dictionary* entries for "Slav" and "Slave," p. 1912, where the two words are seen to derive from each other. But see also Joseph Jacob's *Jewish Contributions to Civilization*, pp. 194-200, where he describes the "wretched business" (p. 200) of supplying slav-slaves to both Byzantium and Persia as a speciality of a group of Jewish traders called Radanites who had a concession with the Khazar kings. Perhaps the terror persisted in folklore.

50. According to Jacobs (see above), the Khazar slav-slave trade flourished until about the year 1,000 A.D. before being curtailed (first) by Duke Sviatoslav of Kiev and (second) by the Mongol invasions.

51. *Columbia Viking Desk Encyclopedia*, entry for "Khazars," p. 565.

52. Howard Fast, *The Jews, The Story of a People* (New York: Dial

Press, 1968), tells of the massive Ashkenazim emigration to America during the latter half of the 19th century.

53. Plenty of Jewish and non-Jewish scholars had attributed the Ashkenazim to the Khazars before 1900. See Koestler's bibliography for *The Thirteenth Tribe*, but note also that Joseph Jacobs, writing in the period of 1890-1920 refers to the Khazar/Jewish connection as a matter of common (scholarly) knowledge.

54. Koestler, *The Thirteenth Tribe*, p. 176.

56. Coon, The Origin of Races, see photo reconstructions, Plate XXXI (Neanderthals) p. 369.

CHAPTER SIX: DELUSIONS, DECEPTIONS AND DISTORTIONS

1. For example, Jacob ben Makhir invented an instrument for measuring angular distances, such as the distance of a star above the horizon, that became known as "Jacob's staff." It was a crude ancestor of the later astrolab and modern sextant, but it greatly assisted medieval European navigation. It may have been inspired by Jacob's knowledge of the Arabic *isbag* method of measuring angles with plaques of established dimensions held at arm's length. See Howard Fast's *The Jews, the Story of a People,* p. 290; and on the same page, Fast briefly mentions the Cresques family's cartography and the navigational tables of Abraham Zacuto.

2. See Bradley's *Holy Grail Across the Atlantic*, pp. 95-109; and *The Columbus Conspiracy,* pp. 119-157. These "portolans" and allied maps have fascinated me for thirty-five years and first inspired me to do historical research—which has led in suprising and unexpected directions.

3. Bradley, *The Columbus Conspiracy*, pp. 212-213.

4. Ibid.

5. Claude Levi-Strauss, *Tristes Tropiques* (New York: Atheneum, 1974), p. 109.

6. Consultation with Dr. Ray Winbugh of the Bishop Joseph Johnson Black Cultural Center, Vanderbilt University, September 21, 1991, relying upon his research into this matter.

7. As above, but I find 100 million victims hard to credit given total world and African population of that time (1450-1850 A.D.) and the registered number of European and North American vessels afloat.

8. Jeffries, *Newsday*, Col. 1, p. 27.

9. Jeffries, *Newsday*, Col. 4, p. 26. Joseph Jacobs in *Jewish Contributions to Civilizations*, p. 244, mentions Aaron Lopez of Newport, Rhode Island as owning "as many as thirty vessels," but giving no totals for vessels owned by the entire Jewish business community of Newport, and no direct statement that Lopez or others were directly involved in the slave trade. However, Jacobs (pp. 218-246) gives a very detailed account of Jewish involvement in the trade of New World mineral and agricultural commodities: precious metals, cotton, rum, molasses, indigo, etc., and the bullion/insurance services that expedited this trade. Although Jacobs refrains from mentioning the "wretched business" of the slave-trade explicitly, two things must be borne in mind: first *all* of Jacobs' above-named New World commodities were plowed back into the slave-trade which allowed these products to be produced. Therefore, while avoiding direct admission that the Jews were involved in the slave-trade from Africa to America, Jacobs gives vast circumstantial evidence that they must have been since the commerce was inter-related.

10. Jeffries, *Newsday*, Col. 2, p. 26.

11. *Emerge* magazine (February 1992), p. 32.

12. Jeffries, *Newsday*, Col. 2, p. 26.

13. Ibid.

14. *Columbia Viking desk Encyclopedia*, p. 894, entry for "pyramids."

15. This is not meant to be facetious. A number of nomadic breads were made in a circular fashion so that they be could be strung on cords or rope for easy carrying. I've seen northern Finnish breads made bagel-fashion for easy carrying, threaded on cords, fastened to reindeer sleds. *The American Heritage Dictionary* entry for *bagel* (p. 54) derives the word from the Old Germanic *boug*, meaning a ring, circle. But it is tempting to go further back and to derive *bagel* from the Indo-European root *bheug* (pronounced like Brooklynese for "berg"=boig), *American Heritage Dictionary of Indo-European Roots*, p. 809, where we find that *bheug* means any swelled or curved object. Now, nomadic breads swelled by yeast and curved to be threaded onto cords or ropes to make for easy carrying, possibly came from this root word in primal Indo-European. It is interesting that the first "luggage" was probably a leather bag that "swelled" and "curved" when it was fillied with things to be carried—hence *baggage* from the same ancient Indo-European root. Or, for that matter,

one of the first people to enter Ireland were the fir-*bolg*, or "men with bags," men who filled leather bags with earth which they moved to make primitive earthworks and hill forts. This ancient Goedelic fir-*bolg* comes from the same primal Indo-European root, bheug, something swelled into a curved shape.

16. It is barely possible that the modern Russian word for a cart-driver, *balagula*, comes from the same from the same Indo-European root word, *bheug*, as explained above; but see Koestler, *The Thirteenth Tribe*, p. 155.

17. Jeffries, *Newsday*, Col. 4., 27.

18. *Columbia Viking Desk Encyclopedia*, entry for "slavery," p. 1009.

19. Jeffries, *Newsday*, Col. 4, p. 27.

20. Ibid.

21. See any of the many pioneering books on revision of African history by Britain's Basil Davidson, as well as many other African-authored books since the 1950's.

22. January 5, 1991.

23. *Emerge* magazine, "Melanin Madness," (February 1992), p. 37.

24. CBS "20/20," (January 12, 1992).

25. Quote from Gen. Colin Powell, "Biography," Arts and Entertainment Television, broadcast in Canada (January 9, 1992).

26. Koestler, *The Thirteenth Tribe*, p. 223.

27. Personal communication to me by Yakov ben Shalom, Israeli fighter pilot and survivor of the "War of Attrition" (he was shot down over Latakia, after helping to develop the "Split S" aerobatic that could evade the Soviet-built SAM missiles); later technical writer for *Canadian Aviation News* (Toronto: Maclean-Hunter) during 1985 where I met him. We collaborated on his war story, "A Hill in Syria," submitted to the Canadian Broadcasting Corporation.

28. I tend to agree with *Emerge* magazine (February, 1992), p. 32 which noted that Dr. Leonard Jeffries assigned to Africa an "artificial and romantic superiority." Some afrocentrists, in trying to redress an unjust Caucasian version of history, have gone overboard in the other direction by claiming everything progressive as deriving from Africa. This is understandable, but regrettable. My good friend, Dr. Ray Winbush of Vanderbilt University, maintains (September 21, 1991, personal conversation) that "Ancient Egypt" was a wholly "Black" achievement. In a

similar vein, Ray Winbush asserted his belief that even the Far Eastern civilizations of Ancient China were inspired by Negroid cultural expertise.

When such opinions are not supported by facts they become merely over-reactions to the White Man's distortions of history. African achievements are so real, and so verifiable, *that they do not need to be exaggerated.* I have tried to stress this in many talks and presentations. Yet, on January 25, 1992 in Chicago at Kennedy King College, I was threatened by some Blacks in the audience because I dared to suggest that Dr. Len Jeffries might have "exaggerated and romanticized" the very real nature of African achievements.